LAST-INSTAR LARVAE AND PUPAE OF THE SIMULIIDAE OF BRITAIN AND IRELAND:

A KEY WITH BRIEF ECOLOGICAL NOTES

WITHDRAWN

by
JON BASS

*Institute of Freshwater Ecology,
River Laboratory*

FRESHWATER BIOLOGICAL ASSOCIATION
SCIENTIFIC PUBLICATION No. 55

1998

Series Editor: J. M. ELLIOTT

Sparsholt College Library
Sparsholt, Winchester
Hants. SO21 2NF

The Environment Agency welcomes the opportunity to part-sponsor the publication of *Last-instar larvae and pupae of the Simuliidae of Britain and Ireland: a key with brief ecological notes,* the latest in the FBA's series of Scientific Publications which continue to provide essential authoritative information needed to underpin the management of freshwater habitats.

The key and associated notes represent an invaluable source reference for identification, distribution and ecology. High quality information is required to maximise the effectiveness of measures to assess and monitor the state of rivers as part of the Environment Agency's aim to achieve a continuing improvement in the water environment in England and Wales. Moreover, accurate identification of rarer macroinvertebrates is essential with respect of the Agency's duty to further conservation.

The Environment Agency strongly recommends this FBA publication to practitioners and enthusiasts alike, but the Agency accepts no responsibility whatsoever for any errors or mis-statements contained in this publication. Any queries or comments should be directed to the FBA.

Published by the Freshwater Biological Association,
The Ferry House, Far Sawrey, Ambleside, Cumbria LA22 0LP

©Freshwater Biological Association 1998

ISBN 0 900386 58 4

ISSN 0367-1887

PREFACE

One of the earliest publications (Number 9) of the Freshwater Biological Association was a key to British Simuliidae by John Smart. This appeared in 1944, but was not replaced until 1968 when a completely revised and updated edition was published as Scientific Publication Number 24. This new key was the end product of the extensive work of Lewis Davies and provided the standard reference work for many years. Since publication of this important monograph, there has been a steady increase in information on blackflies, especially on their larvae. This new key to last-instar larvae and pupae by Jon Bass is therefore most welcome as an up-to-date account of this important family of aquatic insects. It does not include a key to adults but the majority of species in Britain and Ireland can be identified by using the earlier key by Lewis Davies. Note, however, that some of the names have changed and that there are three, newly described, additional species (see Introduction in the present publication).

Some of the reviewers of our recent Scientific Publications have bemoaned the absence of distribution maps. This is a deliberate policy for three reasons. First, the value of such maps is questionable when they are based on a limited number of records. Second, the compilation of detailed maps requires a tremendous effort and would be difficult to execute without the availability of a reliable key. Third, the collation of such records and the preparation of distribution maps is ably done by the Biological Records Centre at the Institute of Terrestrial Ecology, and this is the appropriate body to organise the collection of such information.

We are most grateful to The Environment Agency for sponsoring the publication of this new key which will be of value to those who use these insects as indicators of water quality, and to those with an interest in the ecology of this important family of aquatic insects.

The Ferry House
November 1997

J. M. Elliott
Series Editor

CONTENTS

	PAGE
INTRODUCTION	5
Checklist and taxonomic notes	6
Habitats, substrata and behaviour	11
General morphology	12
Collection	18
Preservation and examination	18
KEY TO THE LAST-INSTAR LARVAE	20
Short-cut key for last-instar larvae	44
KEY TO PUPAE	46
SUPPLEMENTARY NOTES	76
ACKNOWLEDGEMENTS	96
REFERENCES	97
INDEX TO SPECIES	102

INTRODUCTION

The Simuliidae, or blackflies, are a family of highly specialised insects. The aquatic stages – eggs, larvae and pupae – are strongly associated with flowing-water habitats. On emergence as adult flies, the females of many species seek a bloodmeal and in this context the family has achieved a formidable notoriety on a world scale. With some notable exceptions, the blackflies recorded from Britain and Ireland cause little irritation to man as adults and no serious risk of disease transmission. Consequently the majority of descriptive, distributional and particularly ecological studies undertaken in Britain have concentrated on the larval and pupal stages of the life cycle. Since Lewis Davies' extensive work on simuliids in the 1960s there has been a slow but steady accretion of new information, much of which is drawn together within the separate sections of this key. For a comprehensive account of blackflies worldwide, "The Natural History of Blackflies" (Crosskey 1990) is recommended. The majority of species found in Britain and Ireland may be identified as adults using the earlier key published by the Freshwater Biological Association (Davies 1968), unfortunately now long out of print; however, the changes in nomenclature summarised in the checklist below (p. 6) need to be taken into account. Additionally, the more recent records and newly described species should also be considered: *Metacnephia amphora* (Ladle & Bass 1975), *Simulium pseudequinum* (Crosskey 1981) and *Simulium juxtacrenobium* (Bass & Brockhouse 1990).

CHECKLIST AND TAXONOMIC NOTES

The checklist of blackflies of Britain and Ireland compiled by Crosskey (1991) has been followed here, with the exception of one recent revision (Bass, Crosskey & Werner 1995). The list of 32 currently valid names is given below, together with their corresponding names used in the key by Davies (1968). Changes and additions to the fauna are explained in the appended notes (1–21).

Current name	Davies (1968)
Prosimulium hirtipes (Fries)	*P. hirtipes*
Prosimulium latimucro (Enderlein)[1]	*P. inflatum*
Prosimulium tomosvaryi (Enderlein)[2]	*P. arvernense*
Metacnephia amphora Ladle & Bass[3]	(added species)
Subgenus *Hellichiella*	
Simulium latipes (Meigen)[4]	*S. subexcisum*
Subgenus *Nevermannia*	
Simulium angustitarse (Lundström)[5]	*S. cambriense*
Simulium lundstromi (Enderlein)[6]	*S. angustitarse*
(synonymised with *lundstromi*)	*S. latigonium*
Simulium armoricanum Doby & David	*S. armoricanum*
Simulium cryophilum (Rubtsov)[7]	*S. brevicaule*
Simulium vernum Macquart (species-complex)[8]	*S. latipes*
(within the *vernum* species-complex)	*S. naturale*
Simulium juxtacrenobium Bass & Brockhouse[9]	(added species)
Simulium urbanum Davies	*S. urbanum*
Simulium dunfellense Davies	*S. dunfellense*
Simulium costatum Friederichs	*S. costatum*
Subgenus *Eusimulium*	
Simulium velutinum (Santos Abreu)[10]	(unnamed species–'I')
Simulium angustipes Edwards	*S. angustipes*
Simulium aureum Fries	*S. aureum*
Subgenus *Wilhelmia*	
Simulium lineatum (Meigen)[11]	*S. salopiense*
Simulium pseudequinum Séguy[12]	(added species)
Simulium equinum (Linnaeus)[13]	*S. equinum*
(synonymised with *equinum*)	*S. zetlandense*

Subgenus *Boophthora*
   Simulium erythrocephalum (De Geer)   *S. erythrocephalum*

Subgenus *Simulium*
   Simulium intermedium Roubaud (species-complex)[14]   *S. nitidifrons*
   Simulium ornatum Meigen (species-complex)[15]   *S. ornatum*
   Simulium trifasciatum Curtis[16]   *S. spinosum*
   Simulium argyreatum Meigen[17]   *S. monticola*
   Simulium variegatum Meigen   *S. variegatum*
   Simulium tuberosum (Lundström) (species-complex)[18]   *S. tuberosum*
   Simulium rostratum (Lundström)[19]   *S. sublacustre*
   Simulium morsitans Edwards   *S. morsitans*
   Simulium posticatum Meigen[20]   *S. austeni*
   Sirnulium reptans (Linnaeus)   *S. reptans*
   Simulium noelleri Friederichs[21]   *S. argyreatum*

Notes

(1) *Prosimulium latimucro* (Enderlein)
Described previously as *Prosimulium inflatum* Davies (Davies 1957, 1968) before the discovery of an earlier description and associated material giving priority to the specific name *latimucro* (Zwick 1974).

(2) *Prosimulium tomosvaryi* (Enderlein)
Initially described as a variety of *Prosimulium hirtipes* by Grenier (1947), using the name *arvernense*. It was later elevated to species status by Davies (1966). Zwick (1974) examined material from Europe described as *tomosvaryi* by Enderlein and considered this identical to *arvernense* from England. The name *tomosvaryi* therefore takes precedence.

(3) *Metacnephia amphora* Ladle & Bass
First discovered in a seasonally-flowing winterbourne in Dorset (Ladle & Bass 1975). A record of a closely similar species, *Metacnephia tredecimata* (Edwards) and its description, was included in Davies' (1968) key (as *Cnephia tredecimatum*). However, the presence of *tredecimata* within Britain and Ireland has never been confirmed; this lead to its removal from the more recent checklist (Crosskey 1991). A cytotaxonomic study of Palaearctic *Metacnephia* species (Procunier 1982) supported the separation of *amphora* from *tredecimata*.

(4) *Simulium latipes* Meigen
British records of this species were initially as *Simulium subexcisum* (Edwards 1920) and repeated by later authors until synonymy was established with *Simulium latipes* (Crosskey & Davies 1972). Unfortunately this resulted in the transfer of the name *latipes* from a common small stream species (renamed *Simulium vernum* Macquart (Crosskey & Davies 1972)) to a rare species (formerly *subexcisum*) with similar habitat preferences. Post-1972 references to *latipes* should be checked to establish which species has been identified.

(5) *Simulium angustitarse* (Lundström)
Previously recorded as *Simulium cambriense* Davies (Davies 1968) and synonymised by Zwick (1974) with *angustitarse*. Davies (1968) recorded three closely related species: *angustitarse, cambriense* and *latigonium*. *Simulium angustitarse* in Davies' sense was later synonymised with *lundstromi* by Zwick (1974).

(6) *Simulium lundstromi* (Enderlein)
Previously recorded as *Simulium angustitarse* Lundström by Davies (1966, 1968) but synonymised with *lundstromi* by Zwick (1974). *Simulium latigonium* and *lundstromi* are now regarded as the same species (Bass, Crosskey & Werner 1995) and the name *latigonium* should be treated as a synonym of *lundstromi*.

(7) *Simulium cryophilum* Rubtsov
Originally recognised as a component of the British simuliid fauna (as *brevicaule*) by Davies (1966). He considered *Simulium carthusiense* "form *brevicaule*" was a full species and used the name *Simulium brevicaule* Dorier & Grenier. Just prior to this Knoz (1965) had synonymised *brevicaule* with *Simulium cryophilum*.

(8) *Simulium vernum* sl. species-complex
Simulium vernum Macquart was previously recorded as *latipes* by Davies, later corrected by Crosskey & Davies (1972). Recent studies have confirmed the presence of a range of closely related species (a species-complex) within *vernum* and *naturale*. These species are morphologically indistinguishable both as larvae and pupae (aside from chromosome banding features (Brockhouse 1985)).

Specimens identified by Davies as *naturale* were recorded in peaty bog streams in northern England and Scotland, generally during summer months. A re-examination of this material, currently held in the Natural History Museum, failed to confirm the apparently clear-cut distinguishing features between pupae of the *vernum* species-complex and *naturale*. Jensen (1984) recorded what he regarded as *naturale* from four Danish sites, all oligotrophic first-order streams. Seasonal occurrence was not noted.

More recently, cytological examinations have indicated that *naturale*, from one of Davies' collection sites, is indistinguishable from the "Dorset IIs 2+3" *vernum* sibling, while it is distinct from the Danish *naturale* (Brockhouse 1985 and pers. comm.). The Danish *naturale* pupae had lower common stalks on the gill filaments "about twice the length of the upper" (Jensen 1984). Consequently, *Simulium naturale* Davies corresponds to the *vernum* group sibling "Dorset IIs 2+3" of Brockhouse (1985) and morphological separation within the *vernum* species-complex is not possible (for more details see the descriptions of species given in the supplementary notes on pages 80–82).

(9) *Simulium juxtacrenobium* Bass & Brockhouse
A species with close affinities to the *vernum* species-complex (Bass & Brockhouse 1990; Brockhouse 1985). Re-examination of extensive collections in the Natural History Museum suggest it has not been collected and misidentified in the past. Its precise and restricted habitat requirements, with larval, pupal and adult stages only occurring in the late winter and spring (Bass & Brockhouse 1990), are probably the reason why *juxtacrenobium* was previously overlooked.

(10) *Simulium velutinum* (Santos Abreu)
An extensive cytological study (Leonhardt 1985) confirmed the widespread distribution of *angustipes* and *velutinum* throughout Britain (*angustipes* sibling "E" and *velutinum* sibling "I" of Dunbar (1959)). Examination of concurrently obtained larvae and pupae of these two species indicates that distinguishing characters appear to be confined to the adults (Bass 1985).

(11) *Simulium lineatum* (Meigen)
Simulium salopiense Edwards was synonymised with *lineatum* by Crosskey & Davies (1972).

(12) *Simulium pseudequinum* Séguy
This species was recognised as a component of the British fauna by Crosskey (1981). It had previously been mistaken for the closely similar *lineatum* in a range of collections from central southern England.

(13) *Simulium equinum* (Linnaeus)
Now considered to include *Simulium zetlandense* (Crosskey 1991).

(14) *Simulium intermedium* Roubaud (species-complex)
Simulium nitidifrons Edwards was synonymised with *intermedium* by Crosskey (1987b). Post (1980) had distinguished two cytotypes (of *nitidifrons*).

(15) *Simulium ornatum* Meigen (species-complex).
Post (1980) distinguished four cytotypes in England; their status remains uncertain.

(16) *Simulium trifasciatum* Curtis
Simulium spinosum was synonymised with *trifasciatum* by Crosskey (1982).

(17) *Simulium argyreatum* Meigen.
Referred to incorrectly as *Simulium monticola* in earlier publications in Britain (Crosskey 1991).

(18) *Simulium tuberosum* (Lundström) (species-complex)
Studies on chromosome banding have revealed the occurrence of more than one species (Crosskey 1987a).

(19) *Simulium rostratum* (Lundström)
Simulium sublacustre Davies has been synonymised with *rostratum* (Zwick 1978).

(20) *Simulium posticatum* Meigen
Referred to as *Simulium austeni* Edwards prior to being synonymised with *posticatum* (Zwick & Crosskey 1980).

(21) *Simulium noelleri* Friederichs.
Davies referred to this species as *Simulium argyreatum* but the older name, *noelleri*, was reinstated (Zwick & Crosskey 1980).

HABITATS, SUBSTRATA AND BEHAVIOUR

Location of simuliids

The aquatic stages of Simuliidae may be found in a wide range of flowing waters. Larvae do not occur in situations where the current velocity is too slow for their filter-feeding and respiratory requirements. Intermittent streams are colonised, provided that suitable water currents persist long enough for the life cycle to be completed. Mild organic pollution or enrichment is tolerated and possibly favoured by some species, but generally all are absent from grossly polluted watercourses.

Larvae require a comparatively clean substratum on which to secure a pad of silk strands, which is gripped by their posterior hook circlet. Conspicuous bacterial films and easily detached algal growths on stones and plant surfaces restrict larval attachment. In these situations comparatively high population densities are confined to clean, recently trapped debris such as twigs, leaves and the parts of submerged plants undergoing rapid growth.

Studies on mixed species populations have rarely shown clear microhabitat partitioning between species at the same site. However, changes do occur in species composition and utilisation of substrata over a length of river or stream. To avoid the possibility of selectively recording species, all substrata with sufficient flow impinging upon them should be examined.

Pupae occur at larval attachment sites. However, they are often more numerous where the current velocity is less rapid, such as on the downstream side of rocks or towards the base of submerged plants. Where stones provide a suitable substratum, pupae are frequently aggregated on the largest rocks, though this preference is not as conspicuous as that shown by some caddisflies (Trichoptera). Water velocity patterns may change during the sedentary pupal stage and, in conditions of reduced flow, mortalities occur as algae and fine particles collect around the gill filaments, impairing respiration.

Seasonal variation

At the onset of pupation, males outnumber females and early pupae are generally larger than those found towards the end of an emergence peak, when females predominate. In species with several generations each year (multivoltine), larvae and pupae of the overwintering generation attain a greater size than those of subsequent generations. Therefore, only with the extremely small or large species can size be useful to confirm or cast doubt on an identification.

Rates of growth and the duration of larval life are highly variable. Populations living in cold streams and supplied with particulate material of low nutritional value may take several months to reach the last instar stage. By contrast, in the summer some lowland rivers provide conditions in which growth and development proceed rapidly. In the latter case, the duration of the larval stage can be hard to determine from field data where prolonged recruitment from hatching eggs can lead to a wide size range of larvae and the continuous presence of pupae.

Larval behaviour

Larvae maintain position on a suitable substratum using the posterior hook circlet (Fig. 1), the body projecting downstream parallel with the current. The head fans passively and unselectively capture small food particles from the water. If a larva needs to move or is disturbed, it grips the substratum with the proleg (Fig. 1) and mouthparts, releases the posterior hook circlet and moves in a semi-sideways "looping" fashion to another site without being swept away by the flowing water. If violently disturbed, larvae may release themselves and drift in the current, but frequently they remain attached to the substratum by a thread of silk produced by the salivary glands. They are able to climb back up the thread to their previous attachment point, or drift downstream, trailing a sticky thread which aids re-attachment at a new location.

GENERAL MORPHOLOGY

Larvae

Superficially, simuliid larvae have a similar appearance from the time they hatch until they pupate. This is hardly surprising as throughout this period their life style remains the same and it is commonplace to find larvae of all sizes sharing the same substratum and subjected to similar conditions of water flow.

The larva has a fairly rigid chitinous head capsule, but the cylindrical thorax and abdomen are soft, with a thin transparent cuticle. The larvae have a characteristic dumb-bell shape, with the narrowest point of the body just anterior to the midpoint. Their shape distinguishes simuliids from other dipterous larvae and, indeed, from all other aquatic invertebrates. Their size ranges from *ca.* 0.6 mm (lst instar) to 5–11 mm (last instar), with differences between generations and between species. The number of larval instars (stages between moults) can vary in some species between six and seven.

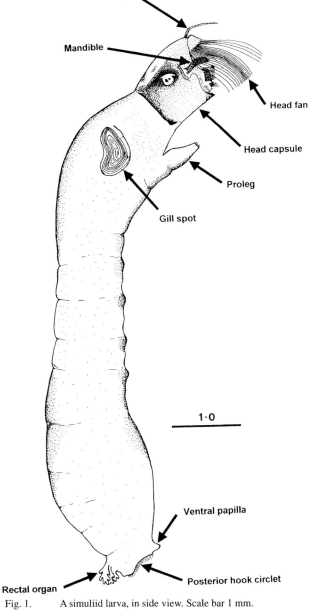
Fig. 1. A simuliid larva, in side view. Scale bar 1 mm.

The head capsule has a conspicuous pair of head fans (cephalic or labral fans) (Fig. 1) used in filter-feeding. Each fan consists of numerous curved rays, resembling combs, bearing fine microtrichia. Preserved larvae may have these fans either spread or withdrawn and tightly closed. The inconspicuous antennae are situated near the base of each head fan (Fig. 1).

Mandibles and a variety of highly modified mouth parts, all of which are infrequently used to aid identification, are situated ventrally. A toothed plate, the hypostomium, lies between the articulating mouth parts in a central anterior position (Fig. 2). Behind this, the rear of the head capsule is indented by the postgenal cleft (Fig. 2). The shape and relative size of the cleft is a diagnostic feature used throughout the key to larvae.

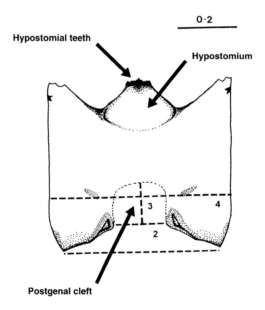

Fig. 2. Ventral view of larval head capsule. Four dimensions used in the key to larvae are indicated by broken lines: 1 = head capsule width; 2 = postgenal cleft width; 3 = postgenal cleft height; 4 = maximum head width. Scale bar 0.2 mm.

Dorsally, the head capsule has two longitudinal sutures in the cuticle, forming the edge of a plate-like area known as the cephalic apotome (Fig. 3). In 1st-instar larvae the apotome has a central pigmented area with a short raised spine, the egg-burster. In later instars the pattern and intensity of apotome marks and the general ground-colour of the head capsule are used in identification. These standard apotome marks are referred to as the postero-median, antero-median, postero-lateral and antero-lateral marks (Fig. 3). Just behind the head capsule, on the thorax near the posterior corners of the apotome, are two small elongated spots – the cervical sclerites (Fig. 3). Paired black eye-spots are situated on each side of the head capsule; generally an eyebrow mark is present in the form of a dark crescent, a pale crescent, or reduced to a single small pigment spot.

The larval thorax has a single appendage, a short proleg (Fig. 1), angled forward and frequently obscuring the postgenal cleft. The proleg bears a terminal circlet of short hook rows. The slight constrictions along the thorax and abdomen correspond to the segmentation of the body. A ventral nerve ganglion may be seen through the cuticle within each segment, and the suboesophagial ganglion often appears dark grey through the postgenal cleft.

The abdomen is thicker than the thorax and external signs of segmentation become less clear posteriorly. Two types of abdominal appendage are visible only on relaxed specimens. Dorsally, a thin-walled rectal organ may be everted. This consists of three lobes, with subsidiary lobules present in many

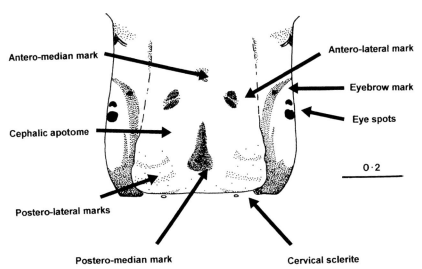

Fig. 3. Dorsal view of larval head capsule. Scale bar 0.2 mm.

species (Fig. 1). Ventrally, a pair of small conical papillae is visible from the side. These papillae consist of the same thin cuticle as the rest of the abdomen and thorax (Fig. 1). The vast majority of preserved larvae have the rectal organ retracted within the rectum, the posterior hook circlet is similarly withdrawn, and the ventral papillae project posteriorly (e.g. Fig. 4). However, rapid fixation or, conversely, slow narcotisation of larvae, can result in a high proportion of larvae with the lobes of the rectal organ extruded and the hook circlet and ventral papillae in their normal position (Fig. 1). In one group of species the ventral papillae are turgid and conspicuous (Figs 1 and 4A); in the other species they are reduced to small folds of cuticle (Fig. 4B).

The penultimate-instar larvae have developing rudimentary pupal gills in the form of pale areas of tissue underlying the cuticle on each side of the thorax, near the proleg (Fig. 1). In last-instar larvae these pupal gill histoblasts, or gill-spots, contain fully formed pupal gills, which become progressively darker as the larvae approach pupation. These spots can be seen by eye in the field and provide a useful indication that last-instar larvae are being collected.

Pupation and pupal morphology

Last-instar larvae construct a tent-shaped cocoon using silk produced by the silk glands (modified salivary glands). The cocoon is firmly attached to the substratum and the tapering posterior end faces upstream. Generally only the extreme anterior part of the pupa, including the gill filaments, protrudes from the open end of the cocoon. The gills arise on each side from a basal trunk, dividing into filaments via common stalks in many species (e.g. Fig. 20). The cocoon may be constructed with differing proportions and patterns of coarse and fine silk strands, depending on the species. Some cocoons possess an anterior projection or horn (e.g. Fig. 27). As the pupa develops, detritus particles and algae frequently coat the cocoon. Initially the pupa is pale, soft and easily damaged. A progressive darkening occurs as maturation proceeds until the fully developed adult can be seen clearly through the pupal skin (exuviae). Shortly before emergence, gas replaces the fluid surrounding the adult within the pupal exuviae.

Characters used extensively in the key to pupae include the following: the shape and texture of the cocoon; the number of gill filaments arising from each side of the thorax; the branching configuration of the gill filaments and width to length ratios of common stalks and basal trunk; the surface texture (microtubercles) of the pupal exuviae in the region of the head and mesothorax.

GENERAL MORPHOLOGY

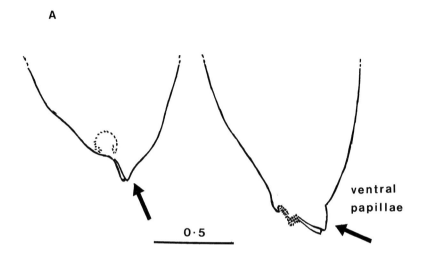

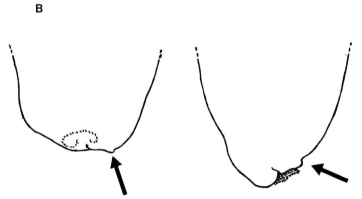

Fig. 4. Ventral papillae (arrowed): in A the papillae are conspicuous, in B they are inconspicuous. Note that drawings were made on larvae with the hind end partially or entirely retracted. Scale bar 0.5 mm.

COLLECTION

As mentioned earlier, larvae and pupae frequently aggregate on clean surfaces and the British species are confined to situations with flowing water. Standard methods of kick sampling and the use of various sampling devices will yield simuliids, though their close attachment to surfaces can result in inefficient recovery rates.

Rocks, stones and plants should be examined carefully for pupae as the close association of larvae and pupae can be helpful in confirming identification. Pupae should be removed with fine forceps to avoid damage. Material may be preserved directly or dark specimens, close to emergence, may be kept alive individually on damp tissue in labelled tubes. Within a few days the adult emerges. After allowing its cuticle to dry and harden it can be preserved with its associated exuviae and cocoon.

Freshly-collected larvae produce copious amounts of silk when placed in a shallow dish of water. With sharp jarring and swirling of the dish the larvae may be dislodged into suspension. They become entangled in the silk threads and can be recovered without the need to preserve, transport and sort through unwanted debris. However, this technique is not suitable for quantitative sampling.

PRESERVATION AND EXAMINATION

Preservation

The efficient fixation, preservation and appropriate storage of specimens is most important to prevent either rapid or long-term deterioration. Two common fixatives* are formaldehyde (4% solution) or ethanol/industrial methylated spirit (IMS) (70–95% aqueous solution); the latter is more pleasant and safer to work with. Both are also suitable as preservatives in the longer term, provided the following precautions are observed.

The fixative/preservative should not be significantly diluted by large volumes of specimens or associated debris (decant and replace with fresh preservative if in doubt); use leak-proof jars/tubes; store in the dark to prevent bleaching; use labels that are unaffected by the preservative.

Protect any specimens that are liable to be shaken in transit by filling the storage container with preservative and introducing some inert packing material (e.g. paper tissue or cotton wool).

---

*Refer to hazard ratings and codes of handling recommended by chemical suppliers.

For long-term storage, submerge containers in larger "bath" vessels, which are in turn sealed, providing insurance against the occasional faulty cap or lid.

Carnoy's Fluid is another very effective preservative, used for examining the banding patterns of the giant polytene chromosomes (recovered from salivary gland cells in last-instar larvae). Suitable larvae are removed directly from their substratum, gently blotted to remove excess water, then plunged into the freshly mixed preservative (3 parts absolute alcohol, 1 part glacial acetic acid). If chromosome preparations are to be made, it is advisable to decant and replace the preservative after about one hour; storage at *ca.* 5°C is also necessary. The advantage of Carnoy's Fluid as a more general preservative is rapid fixation, which results in a high proportion of larvae having a fully everted anal organ and clearly displayed ventral papillae.

Examination

For examining intact specimens at low magnifications (×10 to ×40) a small watchglass or petridish containing 70% IMS is convenient. A powerful lamp providing adjustable top/side lighting is a necessity. Specimens should be handled with fine forceps and manipulated with fine mounted needles.

When temporary slide mounts are required for examination of features at higher magnifications (×100 to ×400), remove the appropriate part using mounted needles and place in an aqueous mountant under a coverslip. It is stressed that whole specimens are too thick for transmitted illumination and the lightly sclerotized parts are easily distorted and displaced when squashed under a coverslip.

For permanent mounts of sclerotized parts, a preliminary overnight immersion in 10% sodium hydroxide solution, followed by thorough rinsing, will aid dissection. Water-miscible mountants are most convenient, though Canada Balsam (preceded by stepped dehydration) remains an alternative.

Examining details of the gill-spot (pupal gill histoblast) (described on page 16) is most efficiently done by detaching the histoblast from the larva and placing it in 50% acetic acid; this encourages the filaments to uncoil slightly, revealing their branching arrangement (×100). Alternatively (at ×10 magnification), tear the overlying larval cuticle with a fine needle and tease out the filaments; care is required as these may become brittle with prolonged fixation.

To examine the surface texture (microtubercles – see pages 16, 55 and 69) of the pupal cuticle, select a dark pupa which will permit easy separation of cuticle from the pupa (pharate adult). Tear away a portion of cuticle from the thorax and head using fine mounted needles, place on a slide under a coverslip and arrange a fold in the cuticle from the thorax area to provide a view of microtubercles in profile.

KEY TO LAST-INSTAR LARVAE

Reference to the width of the postgenal cleft in the larval key refers to the width at the hind margin of the head capsule (Fig. 2), except where it is specified as "width anteriorly". Generally, "head width" also refers to the width at the hind margin of the head capsule (ventral view), unless maximum head width is stipulated, which is slightly further forward in most species. Further details are given in the Supplementary Notes on pages 76 to 96, which should be referred to for confirmation of identification.

On the text-figures, all scale bars are in mm, as indicated.

1 Postgenal cleft short, cleft width exceeds height (arrow, Fig. 5B); cervical sclerites fused to head capsule (arrows, Fig. 5A)— **2**

— Postgenal cleft short or high, if short then width is less than or equal to height (arrow, Fig. 5D); cervical sclerites separate from head capsule (arrows, Fig. 5C)— **4**

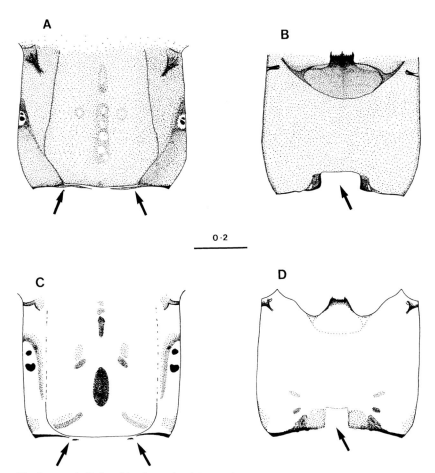

Fig. 5. **A, B:** larval head capsule of *Prosimulium* spp. in A, dorsal view; B, ventral view.
C, D: larval head capsule of *Simulium latipes* in C, dorsal view; D, ventral view.

2 (1) Hypostomium with lateral teeth as long as or more prominent than central tooth (arrow, Fig. 6A). Mandible with numerous long sturdy spines (mandibular serrations) on pre-apical ridge, these longer than gaps between them (arrow, Fig. 6C)—
Prosimulium latimucro (Enderlein)

— Hypostomium with lateral teeth shorter than central tooth (arrow, Fig. 6B). Mandible with pre-apical ridge having numerous short mandibular serrations (arrow, Fig. 6D, E)— 3

3 (2) Pupal gill-spot (Fig. 1) with about 24 coiled filaments. Mandibular serrations decreasing in size and number proximally (arrow, Fig. 6D)— **Prosimulium tomosvaryi** (Enderlein)

— Pupal gill-spot with 16 coiled filaments. Mandibular serrations of similar size and spaced at regular intervals along full length of pre-apical ridge (arrow, Fig. 6E)— **Prosimulium hirtipes** Fries

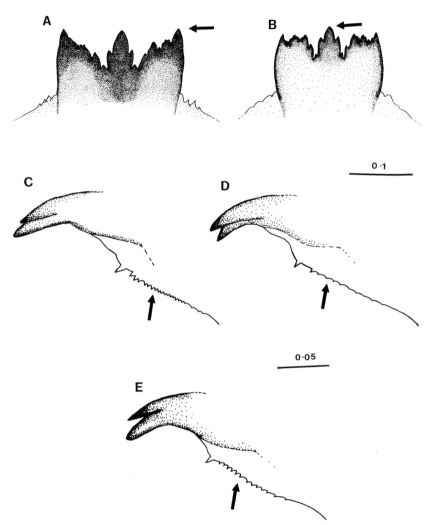

Fig. 6.   **A, B:** hypostomium. A, *Prosimulium latimucro* with central tooth as high as lateral teeth; B, *P. hirtipes/tomosvaryi* with central tooth higher than lateral teeth. **C–E:** mandible. C, pre-apical ridge (arrowed) of *P. latimucro*; D, pre-apical ridge of *P. tomosvaryi;* **E**, pre-apical ridge of *P. hirtipes*.

4 (1) Ventral papillae conical and projecting beyond body of larva in side view (arrows, Figs 1 and 4A). (NOTE: some larvae of *S. erythrocephalum* have conspicuous ventral papillae; larvae with conspicuous papillae *and a postgenal cleft width exceeding one-third the head capsule width* should be taken to couplet 16)— **5**

— Ventral papillae small and inconspicuous (arrows, Fig. 4B)— **15**

5 (4) Antennae dark brown (caution: colour can fade after long storage) with 4–10 pale annulations around the midpoint (arrows, Fig. 7A). Pupal gill-spot with six filaments on three common stalks. Cephalic apotome marks unequally expressed, postero-median mark darkest (Fig. 7B). Head capsule in ventral view as shown in Fig. 7C—
Simulium latipes (Meigen)

— Antennae pale or partially pigmented, without annulations around the mid-point. Pupal gill-spot with only four filaments. Cephalic apotome marks equally expressed— **6**

6 (5) Postgenal cleft posteriorly narrower than one-sixth width of head capsule, cleft narrowing anteriorly. Head capsule ground-colour pale and with ventral pigmentation (if any) restricted to posterior margin (Fig. 8B, D, F) **7**

— Postgenal cleft usually wider than one-fifth width of head capsule (if a little smaller, then cleft sides are parallel). Head capsule ground-colour of uniform pigmentation ventrally, which may be pale to dark (e.g. Figs 10B, 11B)— **9**

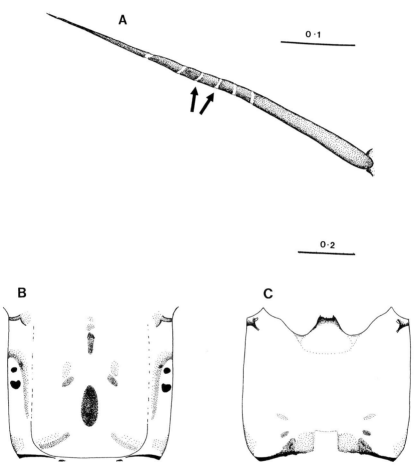

Fig. 7. **A:** larval antenna of *Simulium latipes* with pale annulations. **B, C:** larval head capsule of *S. latipes* in B, dorsal view; C, ventral view.

7 (6) Cephalic apotome with short postero-median mark in form of an equilateral triangle (arrow, Fig. 8A). Postgenal cleft in form of a small inverted "V" (arrow, Fig. 8B). Body pale grey and translucent—
Simulium costatum Friederichs

— Cephalic apotome with postero-median mark in form of a triangle clearly longer than broad (e.g. arrow, Fig. 8C, E). Postgenal cleft with a flat top anteriorly. Body pale but with red-brown annulations— **8**

8 (7) Cephalic apotome marks pale brown (Fig. 8C). No grey smear on head capsule extending from postgenal cleft anteriorly towards hypostomium (arrow, Fig. 8D)— **Simulium angustitarse** (Lundström)

— Cephalic apotome marks medium brown to dark brown (Fig. 8E). Frequently with a grey smear on head capsule extending from postgenal cleft anteriorly towards hypostomium (arrow, Fig. 8F)—
Simulium lundstromi (Enderlein)

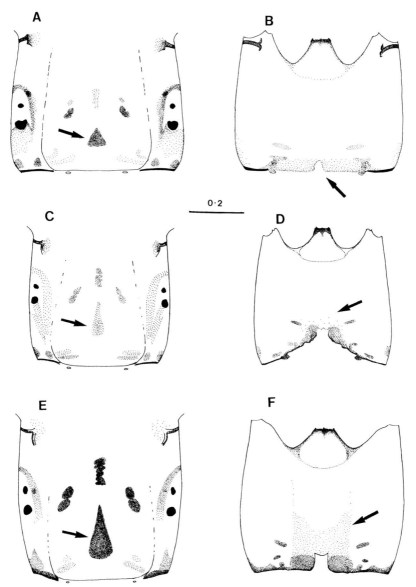

Fig. 8. **A, B:** larval head capsule of *Simulium costatum* in A, dorsal view; B, ventral view.
C, D: larval head capsule of *S. angustitarse* in C, dorsal view; D, ventral view.
E, F: larval head capsule of *S. lundstromi* in E, dorsal view; F, ventral view.

9 (6)   Cephalic apotome postero-median mark in form of a long, narrow rectangle; gap between antero-median and postero-median marks shorter than antero-median mark. Filaments within pupal gill-spot sharply bent and producing a strong antero-ventral angulation (arrow, Fig. 9A)— **10**

—   Cephalic apotome postero-median mark in form of a narrow-based or broad-based triangle, gap between antero-median and postero-median marks equal to or greater than length of antero-median mark. Filaments within pupal gill-spot with rounded outline (arrow, Fig. 9B)— **11**

10 (9)   Cephalic apotome markings dark with fuzzy edges (arrow, Fig. 10A). Postgenal cleft clearly visible against head capsule ground-colour, cleft usually with sharply angled corners anteriorly (arrow, Fig. 10B)— **Simulium aureum** Fries

—   Cephalic apotome markings sharply defined (arrow, Fig. 10C). Postgenal cleft usually with rounded corners, anterior cleft margin contrasts poorly with pale head capsule ground-colour (arrow, Fig. 10D)— **Simulium angustipes** Edwards*
and **Simulium velutinum** (Santos Abreu)*

[*Currently only distinguishable as adults and by features of the chromosome banding pattern in the larvae.]

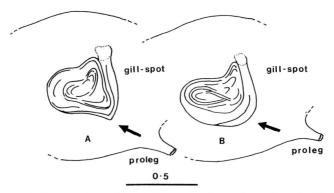

Fig. 9.   **A, B:** side views of larval thorax, showing outlines of pupal gill-spots in A, *Simulium aureum* group; B, other simuliids.

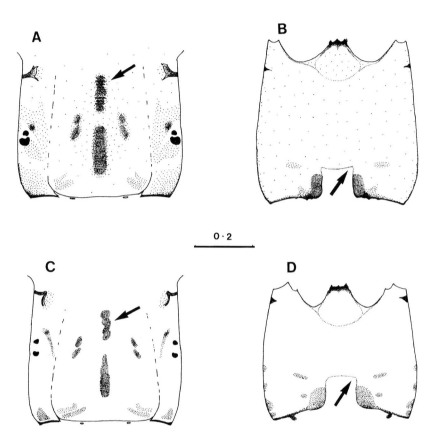

Fig. 10. **A, B:** larval head capsule of *Simulium aureum* in A, dorsal view; B, ventral view. **C, D:** larval head capsule of *S. angustipes/S. velutinum* in C, dorsal view; D, ventral view.

11 (9)   Ground-colour on cephalic apotome paler than rest of head capsule, particularly on lateral areas (arrow, Fig. 11A)*. Postgenal cleft clearly defined, tapering to a point anteriorly, though sometimes rounded (Fig. 11B)—   **Simulium cryophilum** (Rubtsov)

[*Larvae from nutrient-poor habitats and those growing slowly at low temperatures or containing parasites, frequently have darker head capsules or a central grey area on the apotome, the result of extended periods between moults.]

—   Head capsule with uniform ground-colour, e.g. cream (Fig. 11C, D), brown (Fig. 11E, F) or light brown (Fig. 12A, B). Postgenal cleft may or may not be clearly defined—   **12**

12 (11) Postgenal cleft outline contrasting poorly with cream or amber head capsule ground-colour (Fig. 11D). Cephalic apotome with anterolateral marks in form of a pair of clearly defined spots on each side (arrow, Fig. 11C)   **Simulium vernum** Macquart (species-complex)

—   Postgenal cleft well defined, head capsule with light brown or brown ground-colour (Figs 11F, 12B, D). Cephalic apotome with anterolateral marks in a single or double "fuzzy" spot on each side (Figs 11E, 12A, C)—   **13**

13 (12) Head capsule ground-colour brown. Cephalic apotome marks merge with ground-colour of apotome, except for postero-median mark which is darker and with a sharp outline (Fig. 11E). Postgenal cleft small, about equal in width to anterior toothed margin of hypostomium (Fig. 11F)—
           Simulium juxtacrenobium Bass & Brockhouse

—   Head capsule ground-colour light brown. Cephalic apotome marks of uniform colour. Postgenal cleft clearly wider than anterior toothed margin of hypostomium—   **14**

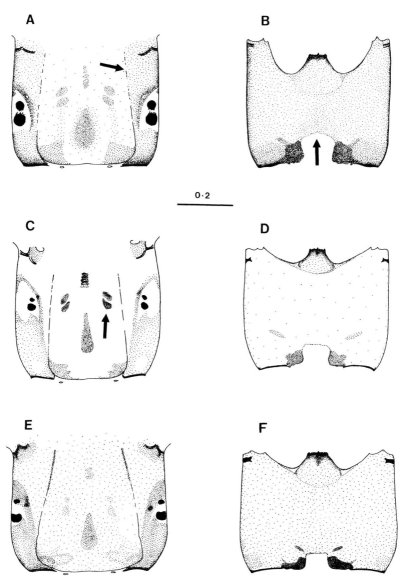

Fig. 11. **A, B:** larval head capsule of *Simulium cryophilum* in A, dorsal view; B, ventral view. **C, D:** larval head capsule of *S. vernum* species-complex in C, dorsal view; D, ventral view. **E, F:** head capsule of *S. juxctacrenobium* in E, dorsal view; F, ventral view.

14 (13) Cephalic apotome postero-median mark a squat, flat-based triangle (arrow, Fig. 12A). Postgenal cleft square with rounded anterior margin (Fig. 12B)— **Simulium armoricanum** Doby & David

— Cephalic apotome postero-median mark longer than broad, with rounded posterior margin (Fig. 12C). Postgenal cleft square or rectangular, with flat, rounded or pointed anterior margin (e.g. Fig. 12D)— **Simulium dunfellense** Davies and **Simulium urbanum** Davies

[Currently, these two species may be distinguished only as adults.]

15 (4) Cephalic apotome with standard apotome marks present and equally pigmented (e.g. Fig. 13A, C)— **16**

— Standard apotome marks unequally pigmented, i.e. antero-median and antero-lateral apotome marks very pale or absent (Fig. 15A)
OR: all standard apotome marks absent (Fig. 15C, E)
OR: standard apotome marks in a "negative" form (Fig. 16C)
OR: standard apotome marks replaced by other pigment patterns (e.g. Fig. 16A)— **19**

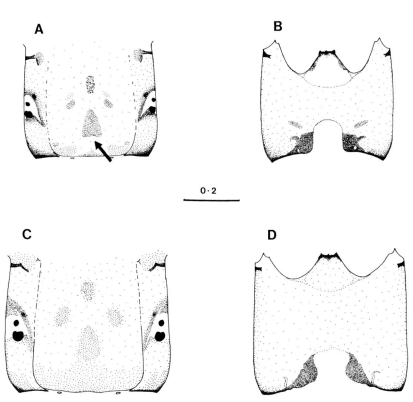

Fig. 12. **A, B:** larval head capsule of *Simulium armoricanum* in A, dorsal view; B, ventral view. **C, D:** larval head capsule of *S. dunfellense/S. urbanum* in C, dorsal view; D, ventral view.

16 (15) Postgenal cleft flask-shaped, with "neck" reaching the base of hypostomial plate (arrow, Fig. 13B). Pupal gill-spot with clearly more than four pairs of filaments when teased out. Cephalic apotome marks condensed, in form of a narrow cross (arrow, Fig. 13A)—
Metacnephia amphora Ladle & Bass

[NOTE: *S. erythrocephalum* larvae collected in early spring may key out at this point, only on the basis of the apotome marks.]

— Postgenal cleft not flask-shaped nor extending to hypostomium base (e.g. Fig. 13D). Pupal gill-spot with either three or four pairs of coiled filaments or convoluted thin-walled lobes, which are not coiled. Cephalic apotome with antero-lateral marks clearly separated from median marks (e.g. arrow, Fig. 13C)— **17**

17 (16) Pupal gill-spot containing broad, thin-walled lobes in form of finger-like structures with convoluted surface texture, lobes not coiled (arrow, Fig. 13E). Cephalic apotome antero-lateral marks clearly paired on each side (arrow, Fig. 13C). Postgenal cleft broad and with ill-defined margin (Fig. 13D)— **Simulium equinum** (Linnaeus)
Simulium lineatum (Meigen)
Simulium pseudequinum Séguy

[By dissection of the pupal gill histoblast, *equinum* may be distinguished from *lineatum* and *pseudequinum* (see the key to pupae, page 64). Generally, these three species are only separable as pupae and adults.]

— Pupal gill-spot with three or four pairs of smooth, thin, coiled filaments (Fig. 9B). Cephalic apotome antero-lateral marks composed of one area of pigment or suffused marks on each side (Fig. 14A, C). Postgenal cleft broad or narrow with ill-defined margin (Fig. 14B, D)— **18**

Fig. 13. **A, B:** larval head capsule of *Metacnephia amphora* in A, dorsal view; B, ventral view. **C, D:** larval head capsule of *Simulium equinum/S. lineatum/S. pseudequinum* in C, dorsal view; D, ventral view. **E:** convoluted filaments within pupal gill-spot on the thorax of larvae of *S. equinum/S. lineatum/S. pseudequinum*.

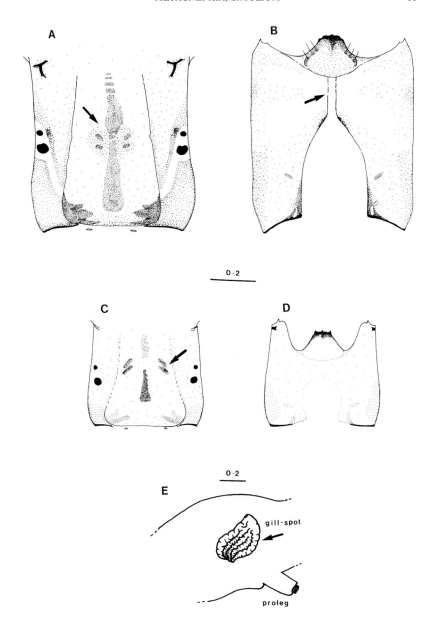

18 (17) Postgenal cleft width at posterior margin about one-third of maximum head capsule width, expanding anteriorly (Fig. 14B). Pupal gill-spot with three pairs of filaments. Cephalic apotome marks darker on spring larvae (illustrated) (Fig. 14A). Abdomen with bands of pigmentation, each abdominal segment with two small inconspicuous papillae dorso-laterally (arrows, Fig. 14E)—
Simulium erythrocephalum (De Geer)

— Postgenal cleft less than one-third as wide as maximum head capsule width (Fig. 14D). Pupal gill-spot with four pairs of filaments. Cephalic apotome markings variable in density throughout the year (Fig. 14C). Abdomen uniform grey without banding, each abdominal segment without papillae dorso-laterally—
Simulium ornatum Meigen (species-complex)
Simulium intermedium Roubaud (species-complex)
and **Simulium trifasciatum** Curtis

[NOTE: Further separation of these species in the larval stage is not possible owing to a high degree of variability, particularly in *ornatum*. This variation is probably exacerbated by the presence of a species-complex in both *ornatum* and *intermedium*.]

Fig. 14.   **A, B:** larval head capsule of *Simulium erythrocephalum* in A, dorsal view; B, ventral view. **C, D:** larval head capsule of *S. ornatum* group in C, dorsal view; D, ventral view. **E:** inconspicuous dorso-lateral papillae on abdominal segments of larva of *S. erythrocephalum* (the head is to the left of the figure).

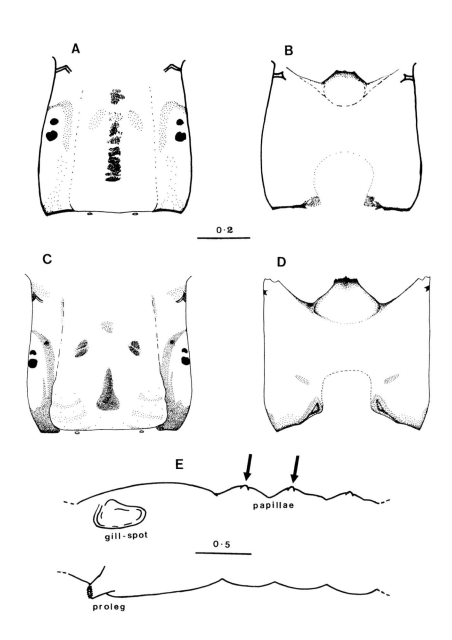

19 (15) Cephalic apotome standard marks pale or absent, though a posteromedian mark may be present. Also an area adjacent to posterior edge of cephalic apotome occasionally pigmented; faint pigmentation may be present in antero-lateral spot positions (Fig. 15A, C, E)— **20**

— Cephalic apotome with standard marks substituted by either pale areas on a dark apotome (Fig. 16C) or heavy pigmentation centrally in form of an "H" (Figs 16A, 17C) or an inverted "U" (Fig. 17A) or a dark central circular mark (Fig. 17E) **22**

20 (19) Postgenal cleft width generally less than one-third of maximum head width, tapering to a point anteriorly. Head capsule ground-colour fawn or brown rendering sides of postgenal cleft clearly visible (arrows, Fig. 15B). Cephalic apotome marks can be restricted to a faint postero-median mark, with other standard marks sometimes present (Fig. 15A)— **Simulium argyreatum** Meigen and **Simulium variegatum** Meigen

[The comparative size of the postgenal cleft and density of head capsule ground colour, previously used to distinguish these species (Davies 1968), has proved unreliable when tested between single-species populations.]

— Postgenal cleft width greater than one-third of maximum head width; cleft margin frequently indistinct (Fig. 15D, F). Head capsule ground-colour pale yellow. Cephalic apotome marks absent or restricted to pale antero-lateral marks and, in some specimens, a pale posterior median mark (Fig. 15C, E)— **21**

21 (20) Paired eye-spots of equal size (Fig. 15C). Postgenal cleft sides parallel, then tapering anteriorly (Fig. 15D). Pupal gill-spot with three pairs of coiled filaments— **Simulium tuberosum** (Lundström)

— Paired eye-spots differing in size, posterior spot larger (Fig. 15E). Postgenal cleft widening anteriorly and with rounded anterior margin (Fig. 15F). Pupal gill-spot with four pairs of coiled filaments— **Simulium reptans** (Linnaeus)

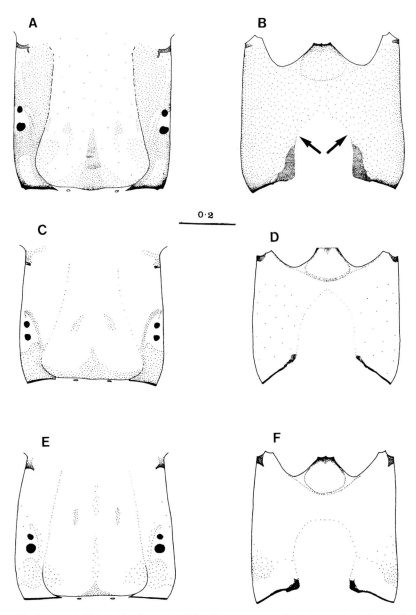

Fig. 15. **A, B:** larval head capsule of *Simulium argyreatum/S. variegatum* in A, dorsal view; B, ventral view. **C, D:** larval head capsule of *S. tuberosum* in C, dorsal view; D, ventral view. **E, F:** larval head capsule of *S. reptans* in E, dorsal view; F, ventral view.

22 (19) Postgenal cleft margin clearly defined by dark head capsule ground-colour (Fig. 16B, D); apotome dark, with or without a dark central mark (Fig. 16A, C)— **23**

— Postgenal cleft margin ill-defined by pale head capsule ground-colour (Figs 15F, 17B, D); apotome pale, with a dark central mark (Fig. 17A, C, E)— **24**

23 (22) Cephalic apotome with an ill-defined central "H"-shaped mark (Fig. 16A). Postgenal cleft tapering to a point anteriorly (Fig. 16B). Pupal gill-spot with eight coiled filaments arranged 3,3,2 or 3,2,1,2—
Simulium noelleri Friederichs

— Cephalic apotome ground-colour brown with pale spots in place of standard markings (Fig. 16C). Postgenal cleft rounded anteriorly (Fig. 16D). Pupal gill-spot with six coiled filaments in three pairs—
Simulium rostratum (Lundström)

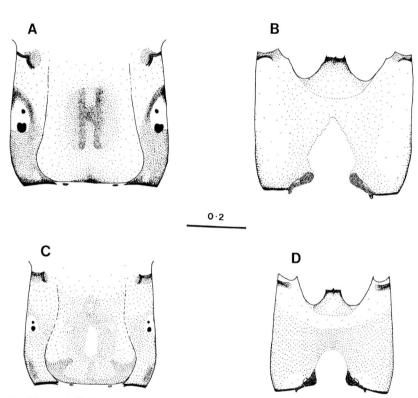

Fig. 16. **A, B**: larval head capsule of *Simulium noelleri* in A, dorsal view; B, ventral view.
C, D: larval head capsule of *S. rostratum* in C, dorsal view; D, ventral view.

24 (22) Cephalic apotome with central mark in form of an inverted "U" (Fig. 17A). Body pale yellow in newly-preserved larvae. Postgenal cleft about one-third as wide as head capsule (Fig. 17B). Pupal gill-spot with six coiled filaments— **Simulium posticatum** Meigen

— Cephalic apotome with central mark in form of a narrow pale brown "H" (Fig. 17C). Body colour not pale yellow. Postgenal cleft less than one-third as wide as head capsule width (Fig. 17D). Pupal gill-spot with four pairs of coiled filaments— **Simulium morsitans** Edwards

— Cephalic apotome with a dark circular central mark (Fig. 17E). Body colour not pale yellow. Postgenal cleft about one-third as wide as head capsule (Fig. 15F). Pupal gill-spot with four pairs of coiled filaments— **Simulium reptans** (Linnaeus)

[This is a form of *reptans* occurring with variable frequency throughout its range; the dark apotome mark is possibly associated with slow larval growth and delayed moulting.]

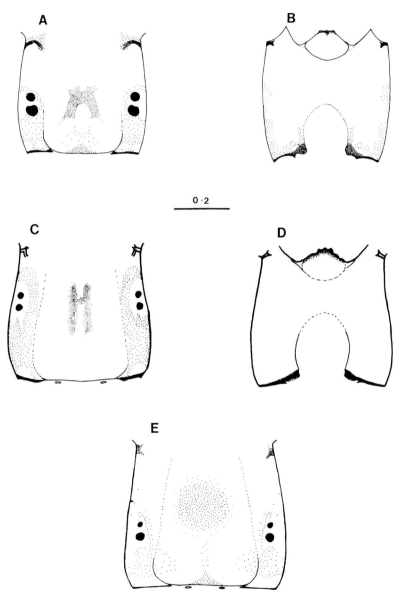

Fig. 17. **A, B:** larval head capsule of *Simulium posticatum* in A, dorsal view; B, ventral view. **C, D:** larval head capsule of *S. morsitans* in C, dorsal view; D, ventral view. **E:** larval head capsule of *S. reptans* (atypical form) in dorsal view.

SHORT-CUT KEY FOR LAST-INSTAR LARVAE

As an alternative to starting at the beginning of the key and working through, it may be possible to save time by taking short cuts. This approach can generally only be adopted when examining a small proportion of last-instar larvae. Material preserved in a relaxed state will have the anal hook circlet fully exposed in the terminal position. Either large conspicuous (Fig. 1) or small vestigial (Fig. 4B) ventral papillae and the presence (Fig. 1) or absence of lobules on the rectal organ lobes need to be clearly seen for these short cuts to work. Species with these various combinations are listed in Table 1. The short-cuts can be used as described below.

1. Conspicuous ventral papillae present* (Fig. 1); rectal organ with or without secondary lobules— **start at couplet 5**
   species in group A, Table 1

   *(A possible exception is *erythrocephalum*; see footnote under Table 1 on p. 45.)

   [The alternative: small fold-like ventral papillae are present on all other species.]

   NOTE: turning to the main key at couplet 5, larvae with this feature
   key out before couplet 15

2. Specimen has both conspicuous ventral papillae and secondary lobules on the rectal organ lobes (Fig. 1)— **start at couplet 11**
   species in group B, Table 1

   NOTE: using the main key, larvae with this combination of characters
   key out before couplet l5

3. Secondary lobules† on the three lobes of the rectal organ (Fig. 1) and small ventral papillae— **start at couplet 20**
   species in group C, Table 1

   †(Secondary lobules on the rectal organ may be absent or vestigial on small larvae of species which show this feature on last-instar larvae.)

4. Larvae with three simple lobes (without secondary lobules) on the rectal organ and small ventral papillae—
   start by checking the postgenal cleft shape (**couplet 1**)
   if *Prosimulium* spp. are ruled out, move to **couplet 15**

   NOTE: larvae with this combination of characters (group D, Table 1)
   key out before couplet 20

Table 1. Species with:-
Group A: conspicuous ventral papillae (Fig. 1, page 13).
Group B: secondary branches (lobules) on rectal organ; conspicuous ventral papillae (Fig. 1).
Group C: secondary branches (lobules) on rectal organ (Fig. 1); small ventral papillae.
Group D: no secondary branches (lobules) on rectal organ; small ventral papillae (Fig. 4B, page 17).

(Table 1 should be used in conjunction with the short-cut key to last-instar larvae and the relevant notes on page 44).

Group A	Group B	Group C	Group D
angustipes	*armoricanum*	*argyreatum*	*equinum*
angustitarse	*cryophilum*	*morsitans*	*erythrocephalum**
armoricanum	*dunfellense*	*noelleri*	*intermedium*
aureum	*juxtacrenobium*	*posticatum*	*lineatum*
costatum	*urbanum*	*reptans*	*ornatum*
cryophilum	*vernum*	*rostratum*	*pseudequinum*
dunfellense		*tuberosum*	*trifasciatum*
juxtacrenobium		*variegatum*	*P. hirtipes*
latipes			*P. latimucro*
lundstromi			*P. tomosvaryi*
urbanum			*M. amphora*
velutinum			
vernum			
erythrocephalum*			

* NOTE: In the case of *erythrocephalum* there may be some difficulty in deciding which form is present. A proportion of small, early-instar larvae have lobules on the lobes of the anal organ. These are absent in the last instar. The ventral papillae of *erythrocephalum* can appear intermediate between conspicuous and vestigial. This species has a large postgenal cleft (ratios of cleft width to head capsule width range from 1:2.5 to 1:3.0). All remaining species in Group D have small postgenal clefts (ratio of cleft width to head capsule width >1:4.25).

KEY TO PUPAE

NOTE:
PUPAE SHOULD BE PARTLY REMOVED FROM THEIR COCOONS
AND EXAMINED IN SIDE VIEW

On the line-drawings of most pupae the cocoon silk thickness and density are illustrated within small inserts, drawn to a scale of 1 mm in width (e.g. Fig. 19). The fine detail of silk threads in these illustrations may not be very clear due to the photographic reduction in size from the original drawings, so selected examples have been reproduced at larger scales in Figs 21 and 22. Some text-figures also carry scale bars, with dimensions in mm as appropriate. Microtubercles on the pupal cuticle range from about 4 µm to 8 µm in different species, and should be examined at ×300 to ×400 magnification (for method of preparation refer to page 19). Gill filament branching characteristics are illustrated in side-view, but lateral deviations of filaments, when they occur in the horizontal plane (e.g. Figs 24 and 26), should be assessed by viewing the pupa head-on. Further information is given in the Supplementary Notes on pages 76 to 96, which should be referred to for confirmation of identification.

1   Each gill consisting of fine filaments (with or without common stalks) arising from a basal trunk (e.g. Fig. 18)—   **2**

—   Gill consisting of short, thin-walled filaments or lobes borne on an inflated basal trunk (Fig. 34)—   **17**

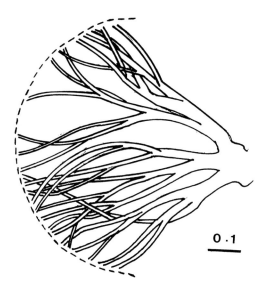

Fig. 18. Gill filaments (truncated in the drawing) from the pupa of *Prosimulium tomosvaryi*.

2 (1) Each gill with more than 20 filaments—   **3**

— Each gill with 13 to 16 filaments—   **4**

— Each gill with 8 filaments—   **6**

— Each gill with 6 filaments—   **11**

— Each gill with 4 filaments—   **19**

    [N.B. Filaments are delicate and prone to damage; check both gills.]

3 (2) Gill with more than 20 filaments (Fig. 18). Cocoon loosely woven, generally covering pupal body—
                       Prosimulium tomosvaryi (Enderlein)

4(2) Gill with about 16 filaments (Fig. 19B). Cocoon consisting of a few silk strands partially covering pupa (Fig. 19A)— **5**

— Gill with 13-15 filaments (usually 14) (Fig. 19E). Well developed and relatively large cocoon enveloping pupal body (Figs 19D, 22)—
Metacnephia amphora Ladle & Bass

5 (4) Gill common stalks diverge sharply from each other (Fig. 19B)—
Prosimulium latimucro (Enderlein)

— Gill common stalks held close together (Fig. 19C)—
Prosimulium hirtipes (Fries)

Fig. 19. **A, B:** pupal cocoon (A) and gill filaments (B) of *Prosimulium latimucro*. **C:** gill filaments of *P. hirtipes*. **D, E:** pupal cocoon (D; also see Fig. 22) and gill filaments (E) of *Metacnephia amphora* (the branching plane of the gill filaments is indicated to the right of the gill).

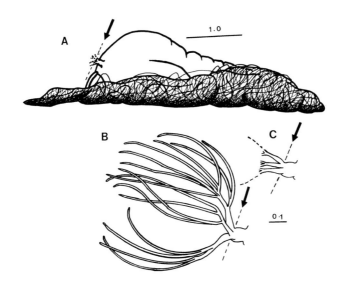
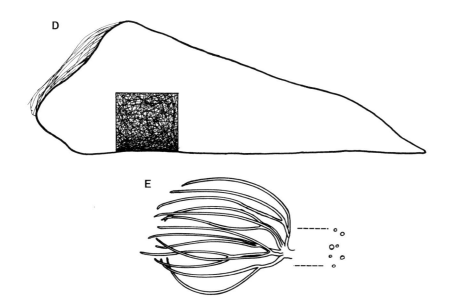

50   KEY TO PUPAE

6 (2) Eight gill filaments arranged 3,3,2 or 3,2,1,2. Cocoon loosely constructed with coarse silk (Figs 20, 22)—
Simulium noelleri Friederichs

— Eight gill filaments in pairs on four common stalks of variable length (e.g. Figs 21, 24). Cocoon of variable texture, loosely or more closely woven, of fine or coarser silk (Figs 22, 23)— **7**

[NOTE: supernumary filaments can occur. If the cocoon bears an anterior process (e.g. Fig. 27), go to couplet 11.]

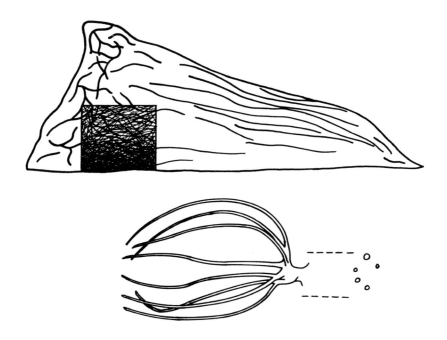

Fig. 20.   Pupal cocoon (also see Fig. 22) and gill filaments of *Simulium noelleri* (the branching plane of the filaments is indicated to the right of the gill).

7 (6) Cocoon with large antero-lateral openings, densely woven from fine silk (Fig. 22). Paired gill filaments on short common stalks (Fig. 21)—
Simulium reptans (Linnaeus)

— Cocoon without antero-lateral openings, texture may be densely or loosely woven from fine or coarse silk. Paired gill filaments on common stalks of variable length— **8**

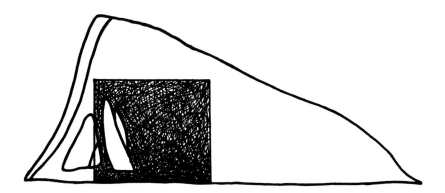

Fig. 21. Pupal cocoon (also see Fig. 22) and gill filaments of *Simulum reptans*.

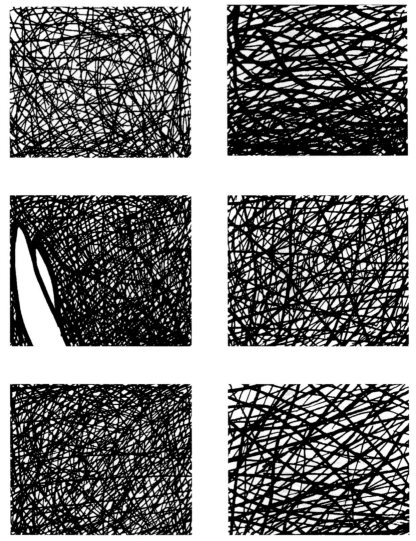

Fig. 22. Some examples of fine and coarse silk threads used to construct cocoons. Top: *Metacnephia amphora* and *Simulium noelleri*. Centre: *S. reptans* and *S. morsitans*. Bottom: *S. ornatum/S. trifasciatum* and *S. intermedium*.

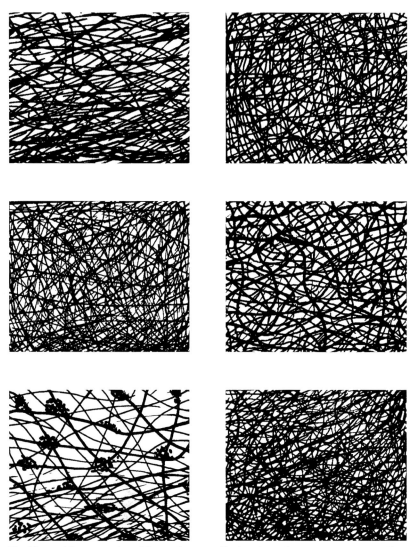

Fig. 23. More examples of fine and coarse silk threads used to construct cocoons. Top: *Simulium posticatum* and *S. costatum*. Centre: *S. vernum* and *S. juxtacrenobium*. Bottom: *S. angustitarse* and *S. lundstromi*.

8 (7) Gill with the two most ventral common stalks turned outwards, their four filaments deviating away from side of the pupa. Cocoon robust, densely woven (Fig. 22), anterior rim slightly thickened (Fig. 24)—
Simulium morsitans Edwards

— Gill with all common stalks directed anteriorly in vertical plane; lowest common stalk may deviate away from side of pupa (Fig. 26). Cocoon robust or less substantial, densely or loosely woven (Figs 22, 23), anterior rim with or without thickening— **9**

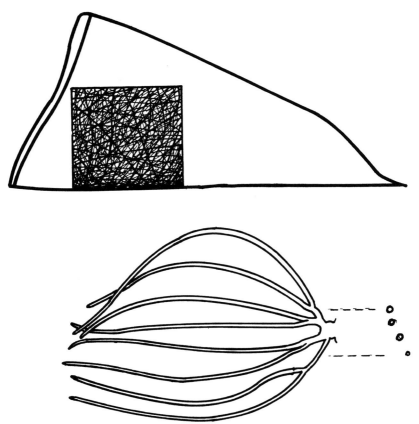

Fig. 24. Pupal cocoon (and see Fig. 22) and gill filaments of *Simulium morsitans* (the branching plane of the filaments is indicated to the right of the gill).

9 (8) Thoracic cuticle antero-laterally with numerous pointed microtubercles (best seen in profile x400) (Fig. 25A)— **Simulium trifasciatum** Curtis

— Thoracic cuticle with rounded and sparsely distributed microtubercles (Fig. 25B)— **10**

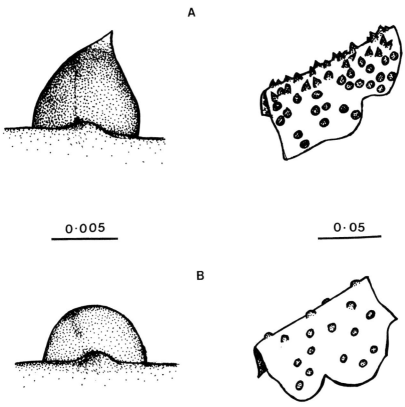

Fig. 25. Microtubercles on thoracic cuticle of: A, *Simulium trifasciatum,* showing numerous pointed microtubercles; B, *S. ornatum/S. intermedium,* showing sparse rounded microtubercles.

10 (9) Cocoon closely woven with fine silk* (Figs 22, 26A)—
Simulium ornatum Meigen (species-complex)

— Cocoon loosely woven with coarse and conspicuous silk threads (Figs 22, 26B)— **Simulium intermedium** Roubaud (species-complex)

[*NOTE: *ornatum* and *intermedium* cannot be distinguished with certainty as pupae, because *ornatum* cocoon silk is occasionally rather thick. Specimens of this type can be mistaken for *intermedium*.]

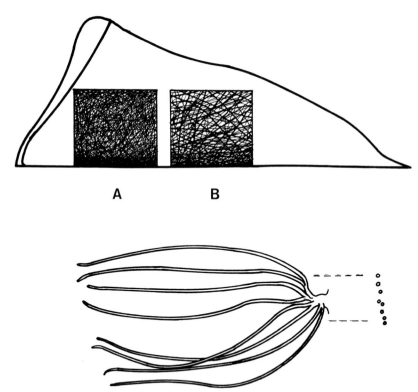

A　　　**B**

Fig. 26.　Pupal cocoon and gill filaments of *Simulium ornatum/S. intermedium/ S. trifasciatum* (the branching plane of the filaments is indicated to the right of the gill). Upper drawing of a cocoon illustrates (in the 1 mm² inserts): A, the fine silk texture of the cocoon of *S. ornatum/S. trifasciatum* (also see Fig. 22); B, the coarser silk texture of the cocoon of *S. intermedium* (also see Fig. 22).

11 (2)   Cocoon with a large anterior horn-like process (Fig. 27)—
            Simulium latipes (Meigen)

   [NOTE: the occasional presence of supernumerary gill filaments has been recorded in this species (Crosskey 1985a). Forms with eight filaments were previously regarded as a separate species (*S. yerberyi*).]

—   Cocoon without such a horn—   **12**

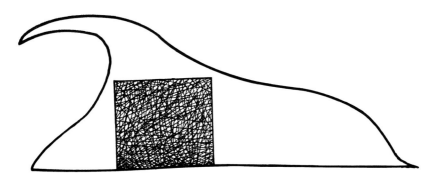

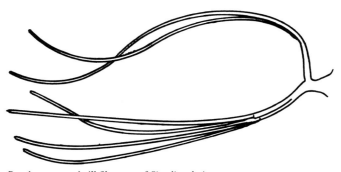

Fig. 27.   Pupal cocoon and gill filaments of *Simulium latipes*.

58 KEY TO PUPAE

12 (11) Gill with paired filaments arising from basal trunk in sequence, lowermost pair first, uppermost pair last (e.g. Fig. 28)— **13**

— Gill filaments, or common stalks, arising together from basal trunk (e.g. Fig. 29)— **14**

13 (12) Thorax of pupa with two oval lumps ("patagia") antero-dorsally (arrow, Fig. 28)— **Simulium variegatum** Meigen

— Thorax of pupa without patagia— **16**

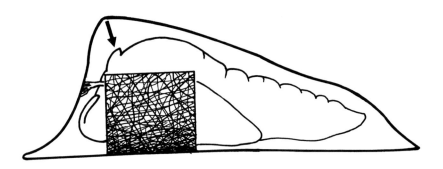

Fig. 28. Cocoon and pupa inside with "patagia", and gill filaments of *Simulium variegatum*.

14 (12) Two central gill filaments deviating in the horizontal plane from one another (arrow, Fig. 29), filaments with very short common stalks, central common stalk frequently absent—
Simulium erythrocephalum (De Geer)

— Gill filaments and short common stalks held in near vertical plane (Figs 30, 31)— **15**

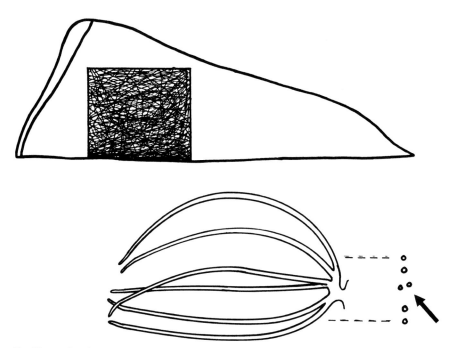

Fig. 29. Pupal cocoon and gill filaments of *Simulium erythrocephalum* (the branching plane of the filaments is indicated to the right of the gill).

15 (14) Gill filaments very slender, remaining well separated at their tips (Fig. 30)— **Simulium rostratum** (Lundström)

— Gill filaments comparatively robust, divergent near base but converging and lying close together at their tips (Fig. 31)— **Simulium posticatum** Meigen

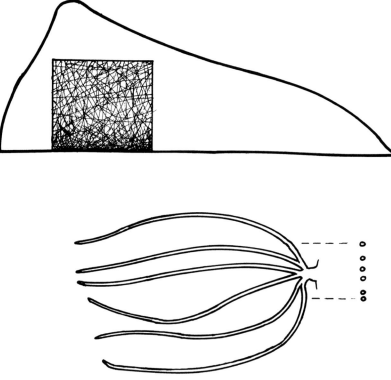

Fig. 30. Pupal cocoon and gill filaments of *Simulium rostratum* (the branching plane of the filaments is indicated to the right of the gill).

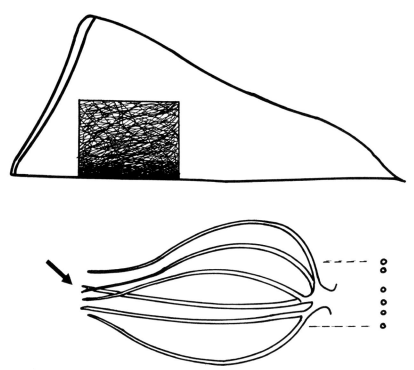

Fig. 31. Pupal cocoon (also see Fig. 23) and gill filaments of *Simulium posticatum* (the branching plane of the filaments is indicated to the right of the gill).

16 (13) Pupa completely covered by cocoon. Gill with uppermost pair of filaments thicker than lower filaments. Cocoon with anterior margin clearly thickened (Fig. 32)— **Simulium argyreatum** Meigen

— Pupa not completely covered by cocoon, anterior part of thorax exposed. Gill with all filaments of similar diameter. Cocoon with anterior margin slightly thickened (Fig. 33)—
Simulium tuberosum (Lundström)

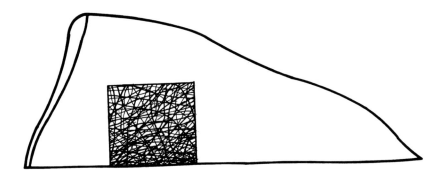

Fig. 32.   Pupal cocoon and gill filaments of *Simulium argyreatum* .

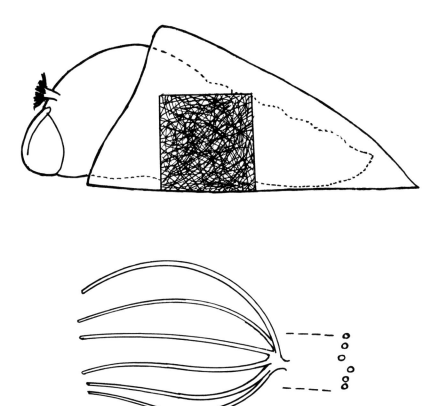

Fig. 33. Pupal cocoon, with thorax of pupa exposed anteriorly, and gill filaments of *Simulium tuberosum* (the branching plane of the filaments is indicated to the right of the gill).

17 (1) Gill with broad thin-walled lobes arising from inflated basal trunk, lobes at their widest point 0.3–0.5 diameter of inflated basal trunk (Fig. 34A)— **Simulium equinum** (Linnaeus)

— Gill with narrow thin-walled lobes arising from inflated basal trunk, lobes much less than half diameter of basal trunk (Fig. 34B, C)—
18

18 (17) Gill lobes without a basal constriction (Fig. 34B)—
Simulium lineatum (Meigen)

— Gill lobes constricted and wrinkled at base where they join inflated basal trunk (arrow, Fig. 34C)— **Simulium pseudequinum** Séguy

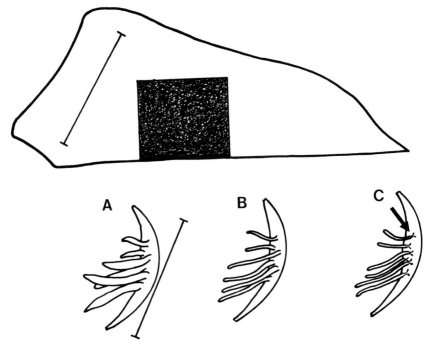

Fig. 34. Pupal cocoon and gill filaments (location of gill within the cocoon is indicated by the scale bar, relative to the 1 mm² inset box) of: A, *Simulium equinum*; B, *S. lineatum*; C, *S. pseudequinum*.

19 (2) Gill with filaments arising in two pairs from two common stalks* (e.g. Figs 35, 36)— **20**

— Gill with two lower filaments lacking a common stalk* (Figs 42, 43)—
26

[*NOTE: some *angustitarse* have lower filaments with a common stalk; compare the cocoon texture with Fig. 23.]

20 (19) Cocoon with anterior rim thickened but without horn-like projection (Fig. 35)— **Simulium costatum** Friederichs

— Cocoon with a horn-like projection on anterior rim (e.g. Fig. 36)— **21**

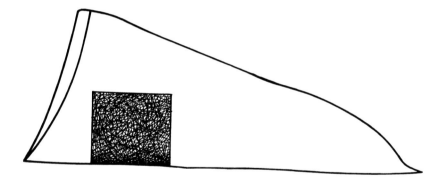

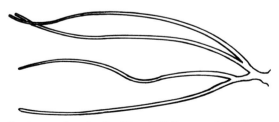

Fig. 35.   Pupal cocoon (see also Fig. 23) and gill filaments of *Simulium costatum*.

21 (20) Gill with lower common stalk clearly longer than upper common stalk (Figs 36, 37)— **22**

— Gill with common stalks about equal in length, or upper common stalk longer than lower common stalk (Figs 38, 40, 41)— **23**

22 (21) Gill filaments diverge in one (vertical) plane near their base (Fig. 36). Cocoon of smooth texture, composed of fine silk (Fig. 23), its horn well developed. Thorax cuticle with sparse low hemispherical microtubercles (mean diameter 4 µm) (Fig. 39A)—
Simulium vernum Macquart (species-complex)

— Gill filaments diverge from one another in two planes at the branching point (arrow, Fig. 37). Cocoon loosely woven, including coarse silk (Fig. 23), its horn well developed or abbreviated. Thoracic microtubercles intermediate in form between Fig. 39A and 39B—
Simulium juxtacrenobium Bass & Brockhouse

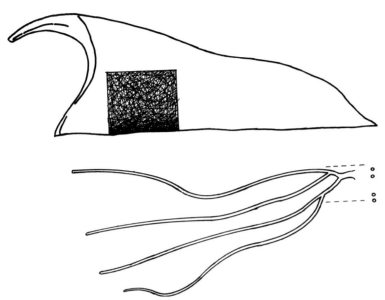

Fig. 36. Pupal cocoon (also see Fig. 23) and gill filaments of the *Simulium vernum* species-complex (the branching plane of the filaments is indicated to the right of the gill).

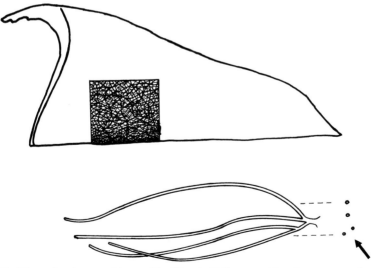

Fig. 37. Pupal cocoon (also see Fig. 23) and gill filaments of *Simulium juxtacrenobium* (the branching plane of the filaments is indicated to the right of the gill).

23 (21) Cocoon pale fawn, finely woven but with lumpy surface (Fig. 38). Gill filaments with long basal trunk, the width to length ratio 1:2.5–1:3.5 (arrows, Fig. 38). Thoracic cuticle with large microtubercles of angular outline (diameter 6–8 µm) (Fig. 39C)—
Simulium armoricanum Doby & David

— Cocoon never of pale colour or lumpy construction. Gill with basal trunk never exceeding a width to length ratio of 1:2.2. Thoracic microtubercles smoothly rounded (Fig. 39B)— **24**

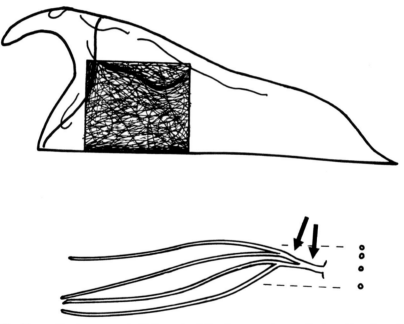

Fig. 38. Pupal cocoon and gill filaments of *Simulium armoricanum* (the branching plane of the filaments is indicated to the right of the gill).

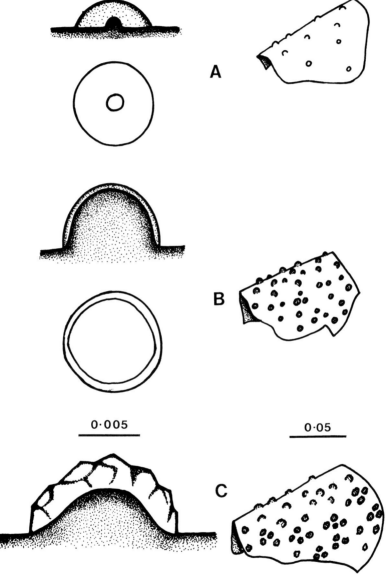

Fig. 39. Microtubercles on the pupal cuticle of: A, the *Simulium vernum* species-complex; B, *S. cryophilum*; C, *S. armoricanum*. Vertical and transverse sections shown at the left. Microtubercle densities shown at the right of the figure.

KEY TO PUPAE

24 (23) Gill filaments held closely together, diverging slightly in the horizontal plane (arrow, Fig. 40). Antennal sheath cuticle with microtubercles— **Simulium cryophilum** Rubstov (typical form)

[*NOTE: *cryophilum* pupae with diverging filaments (atypical form) are occasionally found – see couplet 25.]

— Gill filaments diverging strongly from one another, but sometimes converging towards their tips (Fig. 41A, B). Antennal sheath cuticle with or without microtubercles— **25**

25 (24) Antennal sheath with microtubercles present—
Simulium dunfellense Davies
and **Simulium urbanum** Davies

— Antennal sheath without microtubercles—
Simulium cryophilum Rubtsov (atypical form)

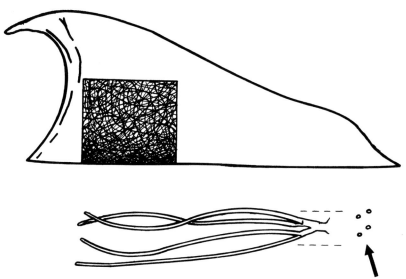

Fig. 40. Pupal cocoon and gill filaments of *Simulium cryophilum* (typical form) (the branching plane of the filaments is indicated to the right of the gill).

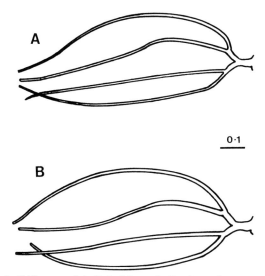

Fig. 41. Pupal gill filaments of: A, *S. dunfellense*; B, *Simulium urbanum*.

26 (19) Cocoon without a horn and evenly woven, its anterior rim thickened (e.g. Fig. 42)— **27**

— Cocoon horn present or absent; if horn absent or reduced to a few silk loops, then cocoon very loosely woven with sparse gaps in the fabric and anterior rim unthickened; if horn present anterior rim often thickened— **28**

27 (26) Uppermost gill filament deviates sharply from other filaments, curving back towards other filaments in a gentle and even arc (arrow, Fig. 42)— **Simulium aureum** Fries

— Uppermost gill filament with distinct (sometimes very sharp) kink not far from the base; uppermost gill filament does not deviate strongly from other filaments (Fig. 43)— **Simulium angustipes** Edwards and **Simulium velutinum** Santos Abreu

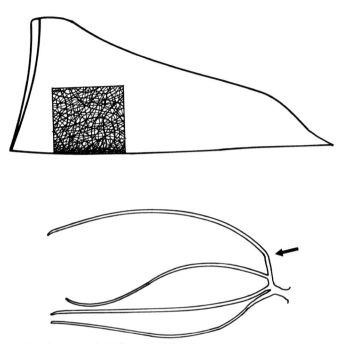

Fig. 42.   Pupal cocoon and gill filaments of *Simulium aureum*.

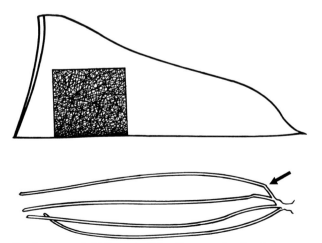

Fig. 43.   Pupal cocoon and gill filaments of *Simulium angustipes/S. velutinum*.

28 (26) Cocoon without a horn, or with only a few silken loops extending anteriorly and loosely woven with perforations (Fig. 23). Gill filaments as long as the cocoon and very slender (Fig. 44)—
Simulium angustitarse (Lundström)

— Cocoon with a conspicuous horn often robustly constructed. Gill filaments short and robust (Fig. 45)—
Simulium lundstromi Enderlein

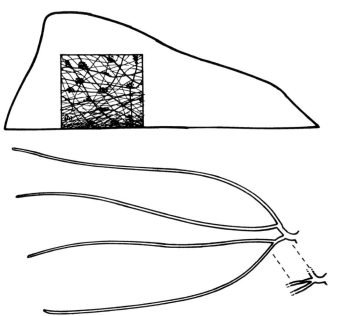

Fig. 44. Pupal cocoon (also see Fig. 23) and gill filaments of *Simulium angustitarse* (variation in common stalk length is indicated to the right of the gill).

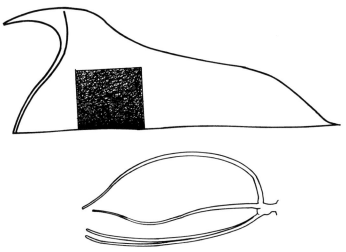

Fig. 45. Pupal cocoon (also see Fig. 23) and gill filaments of *Simulium lundstromi*.

SUPPLEMENTARY NOTES

DESCRIPTIONS OF SPECIES AND BRIEF SUMMARIES OF THE ECOLOGY OF THE AQUATIC STAGES

As larvae and pupae of a single species may vary in size, comparative measurements are generally expressed as proportions in the keys, or direct comparisons are made between the relative sizes (e.g. as ratios) of adjacent structures. The body sizes (length, mm) given below for last-instar larvae of each species include the head, "thorax" and "abdomen", whilst body length for pupae excludes the gill filaments, and cocoon length refers to the whole cocoon.

Combination of characters peculiar to *Prosimulium* species

The following combination of characters is peculiar to the three *Prosimulium* species that are considered separately on pages 76–78.

Larvae. Body colour dark grey or dark brown. Head capsule ground-colour dark brown. Cervical sclerites fused to the sides of the head capsule (Fig. 5A). Postgenal cleft clearly defined, very shallow (width *ca.* ×2 depth) with a flat top (Fig. 5B). Hypostomium with central tooth bearing subsidiary lateral teeth (Fig. 6A, B). Inner pre-apical ridge of mandible with row of numerous short serrations or spines (Fig. 6C, D, E).

Pupae. The cocoon is a loose tangle of silk threads (Fig. 19A), has no distinct shape and may be restricted to an untidy, loose support for the pupal abdomen. Numerous gill filaments, either *ca.* 16 or *ca.* 24 (*tomosvaryi*), on each gill.

Prosimulium hirtipes (Fries)

Larva. Body length 6–8 mm, dark grey. Head capsule ground-colour dark brown; standard apotome marks poorly expressed, may be paler than the ground-colour (Fig. 5A). Postgenal cleft shallow, width clearly exceeding depth (Fig. 5B). Hypostomium lateral teeth shorter than central tooth (Fig. 6B). Inner pre-apical ridge of mandible with a regular row of short spines or serrations, gaps between spines greater than height of spines (Fig. 6E). Ventral papillae small and inconspicuous. Rectal organ with three simple lobes. Pupal gill histoblast with 16 coiled filaments.

Pupa. Cocoon an amorphous loose mass of silk threads covering the pupal abdomen (Fig. 19A). Gill filaments 16; filaments held close together (Fig. 19C).

Ecology. Most authors report *P. hirtipes* occurring in small to large streams in upland areas of Britain, Ireland and mainland Europe (e.g. Davies & Smith 1958; Grunewald 1965; Zwick 1974; Raastad 1975; Carlsson et al. 1977; Zwick & Zwick 1990). It has a single generation of overwintering larvae with pupation occurring in April and May (Davies & Smith 1958; Grunewald 1965; Zwick 1974). Eggs laid in damp moss, above the stream surface, hatch from September onwards (Zwick & Zwick 1990).

Prosimulium latimucro (**Enderlein**)

Larva. Body length 8–10 mm, grey-brown. Head capsule brown; standard apotome marks present (Fig. 5A). Postgenal cleft shallow, width clearly exceeding depth (Fig. 5B). Hypostomium with lateral teeth equal in height to central tooth (Fig. 6A). Mandible pre-apical ridge with narrow gaps between sturdy spines, gaps less than height of spines (Fig. 6C). Ventral papillae small and inconspicuous. Rectal organ with three simple lobes. Pupal gill histoblast with 16 coiled filaments.

Pupa. Cocoon an amorphous, loose tangle of silk threads which rarely covers the pupa (Fig. 19A). Gill filaments 16, on three diverging common stalks (Fig. 19B).

Ecology. *P. latimucro* has a very restricted distribution, recorded from high mountain streams (elevation >600 m) during July to September (as *inflatum*, Davies 1968). Wotton (1976) recorded it from streams in the upper Tees catchment. Recent collections from Gaitscale Gill in the English Lake District yielded last-instar larvae and pupae in April (Bass, unpublished).

Prosimulium tomosvaryi (**Enderlein**)

Larva. Body length 8–10 mm, brown in colour. Head capsule dark brown. Postgenal cleft shallow (Fig. 5B), width clearly exceeding depth. Hypostomium central tooth higher than lateral teeth (Fig. 6B). Mandible pre-apical ridge with spines or serrations reducing in size and number proximally, spine height less than gaps between spines (Fig. 6D). Ventral papillae small and inconspicuous. Rectal organ with three simple lobes. Pupal gill histoblast with *ca.* 24 coiled filaments.

Pupa. Cocoon an amorphous loose tangle of silk threads which may cover the whole pupa. Gill filaments 20–30 (usually 24) on diverging common stalks (Fig. 18).

Ecology. *P. tomosvaryi* (as *arvernense*) has been recorded from a few sites in the north and west of Britain (Doby & Saguez 1963; Davies 1968; Wotton 1976), a single site in Ireland (Fahy 1972) and more recently from six sites on small, iron-rich streams in southeast England (Crosskey, pers. comm.). In mainland Europe its seasonal occurrence is similar to that of *hirtipes* (Grunewald 1965; Zwick 1974), with a single overwintering generation of larvae in the River Fulda, Germany, hatching from eggs laid on moss above the stream surface (Zwick & Zwick 1990). Davies (1968) stated that it was "common" in the English Lake District at sites below 500 m.

Metacnephia amphora Ladle and Bass

Larva. Body length 9–11 mm, tapering anteriorly from the broad hind end, rather than exhibiting the typical dumb-bell shape of other simuliids; body colour medium grey. Head capsule ground-colour fawn or light brown; apotome marks forming a cross-shaped pattern caused by coalescing of the standard marks (Fig. 13A). Postgenal cleft flask-shaped with a narrow "neck" extending to the base of the hypostomial plate (Fig. 13B). Ventral papillae small and inconspicuous. Rectal organ with three simple lobes. Pupal gill histoblast with 14 coiled filaments. Unlike other British species, the cuticle bears a fine micro-sculptured texture of parallel raised ridges or striae spaced 3-4 µm apart (Williams 1980).

Pupa. Cocoon length *ca.* 6 mm; cocoon slipper-shaped, the aperture raised from the substratum (Fig. 19D); silk strands (Fig. 22) frequently obscure the pupa. Gill filaments 14 (Fig. 19E).

Ecology. *M. amphora* was first collected from a temporary chalk stream (a winterbourne) in Dorset (Ladle & Bass 1975); later recorded from similar winterbournes in Dorset, Hampshire and Wiltshire (Bass, unpublished) and West Sussex (Crosskey 1985b). *M. amphora* has a single annual generation, larvae appearing as the winterbournes commence flowing (between November and January). Pupae have been recorded in March and April. Eggs remain dormant in the dry stream bed through the summer months (Cannicott, pers. comm.).

Simulium latipes (Meigen)

Larva. Body length 6–8 mm, pale grey with light brown banding. Head capsule ground-colour pale yellow; standard apotome marks present with postero-median mark darker than the other marks (Fig. 7B). Antennae long, straight, dark brown, with several pale annulations near the mid-point (Fig. 7A). Postgenal cleft small and square in shape; ratios of cleft width to head width range from 1:8.5 to 1:9.0 (Fig. 7C). Conspicuous ventral papillae. Rectal organ with three simple lobes. Pupal gill histoblast with three pairs of coiled filaments.

Pupa. Cocoon length *ca.* 4 mm; cocoon closely woven with thickened anterior rim and large, curved horn (Fig. 27). Gill filaments six; paired filaments arise from three common stalks, upper common stalk curves upwards, its filaments following the profile of the cocoon horn, lower filaments remain close to the substratum (Fig. 27). Occasionally extra filaments are present, arising some distance from the normal common stalk branching points (Crosskey 1985a).

Ecology. *S. latipes* (as *subexcisum*) has been recorded from small, temporary streams draining shallow ponds (Edwards 1920; Davies 1968). More recently *latipes* has been found at a single site in Norfolk (Post 1981), a few sites in southeast England (Crosskey 1985b and pers. comm.) and also a site in southwest Ireland (Bass 1990). *S. latipes* is a rare but widespread species restricted to a single spring generation by its choice of habitat.

Simulium angustitarse (Lundström)

Larva. Body length 6–8 mm, pale grey with reddish-brown annulations. Head capsule ground-colour pale yellow; standard apotome marks pale brown (Fig. 8C). Postgenal cleft small, V-shaped, truncated or occasionally rounded, well defined by pigmentation restricted to the posterior edge of the head capsule (Fig. 8D); ratio of maximum cleft width to head width *ca.* 1:5. Conspicuous ventral papillae. Rectal organ with three simple lobes. Pupal gill histoblast with four coiled filaments.

Pupa. Cocoon length 3.0–3.5 mm; cocoon loosely woven and of thin construction with small holes (Fig. 23), anterior margin without a horn; surface of cocoon frequently coated with detritus particles. Gill filaments four; two pairs of long filaments arising either both from short common stalks or the lower two directly from the basal trunk. Filaments deviating from one another and of uniform diameter (Fig. 44).

Ecology. Davies (1966) first reported *S. angustitarse* (as *cambriense*) from a small stream in Wales; Fahy (1972) recorded it at a single stream site in Ireland. More recently it has been found in streams in Dorset (Ladle & Esmat 1972), Norfolk (Post 1981), on the Isle of Wight and at a single site in the London area (Crosskey 1982, 1985b). It has been recorded throughout central and northern Europe (Zwick 1978). Zwick (1974) found pupae in April and August/September.

Simulum lundstromi (Enderlein)

Larva. Body length 7–9 mm, with reddish-brown bands. Head capsule pale cream with standard apotome marks medium brown to dark brown and sharply defined (Fig. 8E). Postgenal cleft small, tapering or parallel sides, with a flat top; head capsule anterior to cleft frequently with a grey smear of pigment (Fig. 8F); ratios of cleft width to head width range from 1:7 to 1:9. Conspicuous ventral papillae. Rectal organ with three simple lobes. Pupal gill histoblast with four coiled filaments.

Pupa. Cocoon length *ca.* 4.5 mm; cocoon texture densely woven silk (Fig. 23); conspicuous horn present. Gill filaments four; filaments short and robust, upper pair thicker and with a short common stalk, lower pair arising directly from basal trunk; top filament curves upwards, second filament frequently thicker at the base and curves laterally (outwards); all filaments tend to lie close together at their tips (Fig. 45).

Ecology. *S. lundstromi* is recorded from weedy streams and rivers, generally, but not exclusively, in lowland areas of the British Isles. Larvae and pupae attach to vegetation in the main flow, where they are frequently outnumbered by more common species. Pupae can be found throughout much of the year where flow conditions remain suitable, indicating it is multivoltine (Bass et al. 1995).

Simulium vernum Macquart [species-group]: general comments

Davies (1966) described a range of closely similar species which he recognised as distinct from *S. latipes* (now *vernum*; Crosskey & Davies 1972). Earlier studies (e.g. Edwards 1920; Puri 1925; Smart 1944; Zahar 1951; Maitland & Penney 1967) had not distinguished between species with a horn on the cocoon and four gill filaments arranged in pairs, such as *armoricanum, cryophilum, dunfellense* and *urbanum*. Therefore, records of distribution and ecology of "*vernum*" (referred to as *latipes*, before 1972) have to be treated with caution. This is illustrated by Edwards' (1920) observation that there was some variation among "*vernum*" from different sites, noting distinct reddish-brown bands on larvae from the New Forest which were "usually hardly perceptible in Hertfordshire specimens".

Authors are in general agreement that *vernum*-group larvae appear in late winter with a pupal peak in late spring. Further irregular occurrence has been reported in the summer months (Davies 1966, 1968; Maitland & Penny 1967; Ladle & Esmat 1972; Zwick 1974; Post 1981). Wotton (1976) recorded a higher frequency of pupae during summer in the northern Pennines, while Crosskey (1982, 1985b) regarded *vernum*-group species as essentially univoltine in the south of England.

It is noteworthy that Davies distinguished *naturale* from *vernum* (as *latipes*) chiefly on the basis of size (smaller during all life stages), recording a pupal gill common stalk length ratio of greater than 1:2 (ratio of upper to lower common stalk) and a slightly larger larval postgenal cleft. A re-examination of *naturale* collected by Davies and currently held in the Natural History Museum (formally the British Museum (Natural History)) failed to confirm the apparently clear-cut distinguishing features between pupae of the *vernum* species-complex and *naturale*. All pupae in the collection, except one specimen labelled as *naturale*, had upper to lower common stalk length ratios of less than 1:2, compared with the "2–3 times in length" quoted by Davies (1966). Within the *vernum*-complex this ratio is quite variable but rarely falls outside the range of 1:1.0 to 1:2.2 (Bass, unpublished).

Simulium vernum-complex or species-group

The complex contains four sibling species recorded from Britain ("Knebworth", "Dorset IIs–1", "Lymington", "Dorset IIs–2+3") and distinguished cytologically by Brockhouse (1985). These are generally inseparable using conventional morphological features; a broad general description of the complex is given below.

Larvae. Body length 5–7 mm, pale grey with variably expressed red-brown banding. Head capsule ground-colour pale yellow; apotome marks medium brown, antero-lateral marks clearly paired on each side (Fig. 11C). Postgenal cleft small, rectangular, with a flat top or slightly rounded (Fig. 11D); ratios of cleft width to head width range from 1:5.0 to 1:5.5. Conspicuous ventral papillae. Rectal organ with conspicuous secondary branches (lobules). Pupal gill histoblast with four coiled filaments.

Pupae. Cocoon length 3.5–4.5 mm; cocoon closely woven with fine silk (Fig. 23), thickened anteriorly; horn present, usually large and tending to have parallel sides. Gill filaments four; paired filaments arising from two common stalks, the lower common stalk almost always longer than the upper common stalk (ratios within the range 1:1.0 to 1:2.2); filaments branching in the vertical plane (Fig. 36). Sparse, low microtubercles present on the cuticle in the region of the head (excluding antennal sheath) and thorax (Fig. 39A).

Ecology. There are small morphological variations between populations of *S. "vernum"* occupying different streams, but these could be attributable to food supply, water chemistry and temperature effects. However, certain environmental conditions are associated with the occurrence of particular sibling species (Brockhouse 1985). Studies of polytene chromosomes from larval salivary glands have established that at least four *vernum* sibling species occur in Britain (Brockhouse 1985), apparently indistinguishable on the basis of gross morphology. These four are listed below.

"Knebworth" – currently the most widespread sibling recorded from southern England, Ireland, mainland Europe and Alaska (Brockhouse 1985 and pers. comm.). In England and Ireland this sibling occurs in partially wooded streams draining clay soils with a neutral pH (Crosskey 1982; Bass, unpublished).

"Dorset IIs–1" – confined to weedy chalk streams with a pH range of 7.5 to 8.3. To date, populations have been found in Dorset and West Sussex (at the latter site co-occurring with "Knebworth" (Brockhouse, pers. comm.)).

"Lymington" – found throughout the New Forest in central southern England, with populations in small acid streams (pH 4.5–6.5) in open and partially wooded areas (Brockhouse 1985).

"Dorset IIs–2+3" – found in acid streams (pH 4.0–5.6) draining open heathland (Brockhouse 1985). The small size of this sibling prompted repeat collections from Davies' sites in Cumbria which had yielded *naturale*. Specimens from a small moorland stream near Coniston proved to have the same diagnostic chromosome inversions as the "Dorset IIs–2+3" *vernum* sibling (Brockhouse, pers comm.).

Associated pupae and larvae of all *vernum* siblings, distinguished cytologically, have been examined and morphometric data indicate there is a continuum of slight differences within and between siblings. Therefore, no reliable separation of *vernum* siblings based on gross morphology is currently possible.

Simulium armoricanum Doby and David

Larva. Body length 4–6 mm, pale brown. Head capsule ground-colour amber; standard apotome marks a uniform pale brown with postero-median mark a squat, flat-based triangle (Fig. 12A). Postgenal cleft square with rounded corners (Fig. 12B); ratios of cleft width to head width range from *ca.* 1:4.5 to 1:5.0. Conspicous ventral papillae. Rectal organ with secondary branches (lobules). Pupal gill histstoblas contains four coiled filaments.

Pupa. Cocoon length 3–4 mm; cocoon pale fawn with well developed horn; cocoon texture finely woven but with a lumpy surface. Gill filaments four in two pairs; long common stalks and particularly long basal trunk (width to length ratios in the range 1:2.5 to 1:3.5) (Fig. 38); microtubercles of the pupal cuticle with rough surfaces (Fig. 39C).

Ecology. *S. armoricanum* is locally common (Dartmoor: Doby & Saguez 1963) but generally restricted to upland and hill streams in the north and west of Britain, with a few records from the East Sussex Weald and Ireland (Crosskey, pers. comm.). Probably two or three generations, with pupae found from May to August (Davies 1968) and in December in southwest Ireland (Bass, unpublished).

Simulium cryophilum Rubstov

Larva. Body length 6–8 mm, pale brown. Head capsule ground-colour pale to medium brown ventrally and laterally, apotome paler than surrounding head capsule (Fig. 11A). Antennae long, brown and strongly curved. Standard apotome marks pale brown with ill-defined edges. Larvae from nutrient-poor habitats and those growing slowly at low temperatures, or containing parasites, frequently have darker head capsules or a central grey area on the apotome, the result of extended periods between moults. Postgenal cleft well defined, usually tapering to a point anteriorly, sometimes rounded (Fig. 11B); ratio of cleft maximum width to head width *ca.* 1:4.5. Ventral papillae conspicuous. Rectal organ with secondary branches (lobules). Pupal gill histoblast with four coiled filaments.

Pupa. Cocoon length 3.5–4.0 mm; cocoon thickly woven; broad-based horn present. Given the range of slight intra-specific variation, cocoons of *cryophilum, dunfellense* and *urbanum* appear indistinguishable. Gill filaments four; paired filaments with short common stalks (shorter than the basal trunk), upper common stalk slightly longer or of equal length to the lower common stalk; filaments diverge to a small extent in vertical and horizontal planes*

(Fig. 40). Numerous hemispherical microtubercles on the cuticle of the head and thorax, absent from the antennal sheath (Fig. 39B).

Typical form: gill filaments remain close together for most of their length.

Atypical form: gill filaments diverge from one another after branching in the horizontal plane. The atypical form has been recorded from a few sites in southern England (Natural History Museum collection: Bass, unpublished) and Ireland (Bass, unpublished) whilst the typical form is more frequently found where both forms coexist.

Ecology. Davies (1968) gives relatively few records for *S. cryophilum* (as *brevicaule*) in southeast England compared with records for *vernum* (as *latipes*), another common small stream species. Recent collections in southeast England have confirmed this geographical trend (Crosskey 1985b and pers. comm.). In Upper Teesdale in northern England, Wotton (1976) recorded *cryophilum* (as *brevicaule*) and *vernum* (as *latipes*) occurring with almost equal frequency. Jensen (1984) associated *cryophilum* with small woodland streams in Denmark. In County Kerry, Ireland, *cryophilum* was the most widely recorded species in small streams on open moorland during early spring (Bass 1990). Although most frequent in spring, larvae and pupae of *cryophilum* can be found throughout the summer months, with possibly two or three annual generations. Currently, there are not enough data to gauge whether the "typical" and "atypical" forms of *cryophilum* have differing ecological requirements.

Simulium juxtacrenobium Bass and Brockhouse

Larva. Body length 6–7 mm, grey-brown. Head capsule light brown to medium brown; apotome marks ill-defined apart from postero-median mark, a sharp-edged narrow triangle with convex base (Fig. 11E); antero-lateral margins of apotome with dark pigment line (this fades after long storage). Postgenal cleft small and slightly wider than long, almost equal in width to the anterior margin of the hypostomium (Fig. 11F); ratios of cleft width to head width range from 1:6 to 1:7. Conspicuous ventral papillae. Rectal organ with secondary branches (lobules). Pupal gill histoblast with four coiled filaments.

Pupa. Cocoon length *ca.* 4.5 mm; cocoon woven with predominantly coarse silk strands (Fig. 23), cocoon aperture meets the substratum anteriorly at an angle of 70-85° (as seen in *cryophilum, dunfellense* and *urbanum*; in the *vernum*-complex this angle is about 50°); horn usually well developed, sometimes abbreviated. Gill filaments four; filaments deviate in two planes, upper pair on a very short common stalk deviate widely in the vertical plane, lower pair on a longer common stalk (×2.0 to ×3.5 upper common stalk length) deviate slightly from one another in the horizontal plane (Fig. 37).

Ecology. A recently described species having close similarities with *Simulium crenobium* Knoz from eastern Europe and the *vernum*-complex, *S. juxtacrenobium* has been recorded from small acid (pH 3.2–5.5) streams in the New Forest and similar streams draining open moorland in southeast Dorset and southwest Ireland (Bass & Brockhouse 1990). It has a single generation in early spring, with larvae recorded from February to April and pupae from April to May.

Simulium urbanum Davies and *Simulium dunfellense* Davies

Larvae. In both species, body length 6–9 mm, grey-brown. Head capsule ground-colour a uniform pale brown; apotome marks medium brown, frequently with fuzzy boundaries (Fig. 12C). Antennae long, brown and strongly curved. Postgenal cleft square or rectangular, cleft clearly visible against the pale brown ground-colour; anterior margin generally rounded at corners but sometimes flat or pointed (Fig. 12D); ratio of cleft width to head capsule width *ca.* 1:4.5. Conspicuous ventral papillae. Rectal organ with secondary branches (lobules). Pupal gill histoblast with four coiled filaments.

In all larval characters examined, *urbanum* appears to be indistinguishable from *dunfellense*. Davies (1966) regarded *urbanum* as a larger species than *dunfellense*. However, larvae, pupae and adults of *dunfellense* collected from the English Lake District during spring were of comparable size to *urbanum* (Bass, unpublished).

Pupae. In both species, cocoon length 3.5–4.0 mm; cocoon with coarse and fine silk strands; horn generally short but broad-based and always present. Given the range of slight intra-specific variation, cocoons of *cryophilum*, *dunfellense* and *urbanum* appear to be indistinguishable (Fig. 40). Gill filaments four, arranged in two pairs; common stalks and basal trunk are of similar lengths; filaments diverge widely in the vertical plane but curve towards each other at their tips (Fig. 41A, B). Microtubercles of the pupal cuticle extend onto the antennal sheath; microtubercles hemispherical in shape (i.e. taller than those of the *vernum*-complex).

Separation of *urbanum* and *dunfellense* has been hindered by the discovery of a spring generation of *dunfellense* (English Lake District – Bass, unpublished) with last-instar larvae and pupae of comparable size to *urbanum*.

The comparatively greater robustness of *urbanum* gill filaments (Fig. 41A), when compared with those of *dunfellense* (Fig. 41B), remains a useful distinguishing character for pupae in summer.

Ecology. *S. urbanum* is recognised as a distinct form from other members of the *vernum*-group (Davies 1966), mainly on the basis of pupal morphology. All stages closely resemble *dunfellense*. Davies (1966) recorded *urbanum* from a few sites in southern England and Fahy (1972) found it at three stream sites in Ireland. Its presence on Stanmore Common in Outer London (type locality) in small ditch-like seepages that run dry in summer, was confirmed in a more recent study of species distribution in the London area (Crosskey 1985b). In Europe it has been reported from small deciduous woodland streams in Denmark (Jensen 1984). Recent collections of simuliids in the New Forest, Hampshire, provided examples of all life stages of *urbanum* in spring (Bass & Brockhouse 1990; Bass, unpublished).

S. dunfellense was first described by Davies (1966) from cold, stony mountain streams in northern England during June and September. Crosskey (1991) reported it from Ireland. Davies concluded that the species was probably univoltine. However, more recent collections of simuliids in high streams in the English Lake District provided examples of *dunfellense* larvae and pupae in April (Bass, unpublished).

Simulium costatum Friederichs

Larva. Body length 8–10 mm, pale grey, particularly translucent before preservation, with the gut contents clearly visible. Head capsule pale yellow; standard apotome marks light brown with postero-median mark an equilateral triangle (Fig. 8A). Postgenal cleft a small inverted "V" (Fig. 8B); ratios of cleft width to head width range from 1:15 to 1:20. Conspicuous ventral papillae. Rectal organ with three simple lobes. Pupal gill histoblast with four coiled filaments.

Pupa. Cocoon length 4.5–5.0 mm; cocoon closely woven of fine silk (Fig. 23), thickened at the anterior rim; no horn present. Gill filaments four; two pairs arise from common stalks of equal length; filaments deviate from one another in the vertical plane (Fig. 35).

Ecology. *S. costatum* was first recorded in Britain by Edwards (1927) and is widely distributed in central Europe, though unrecorded from Ireland (Zwick 1978; Crosskey 1991). All authors report its close association with springs at stream sources (Davies 1968; Niesiolowski 1971; Crosskey 1982, 1985b; Jensen 1984). It is rarely found more than 100 m below the stream source and is generally associated with permanent springs issuing from chalk or limestone strata. Two or three annual generations have been reported (Niesiolowski 1971; Zwick 1974).

Simulium angustipes Edwards and *Simulium velutinum* (Santos Abreu)

Larvae. In both species, body length 6–8 mm, pale with pale brown bands. Head capsule ground-colour pale yellow; eyebrow mark consists of a single small pigment spot; apotome marks pale brown with sharply defined edges, gap between antero-median and postero-median marks about half the length of anterior mark (Fig. 10C). Postgenal cleft square with rounded corners, (hard to see against the pale head capsule) (Fig. 10D); ratio of cleft width to head width *ca.* 1:5. Conspicuous ventral papillae. Rectal organ with three simple lobes. Pupal gill histoblast with four coiled filaments; histoblast outline with a sharply angled corner antero-ventrally (Fig. 9A).

These two species cannot be distinguished in the larval stages at present except by features of the chromosomal banding pattern.

Pupae. In both species, cocoon length *ca.* 4 mm; finely woven of even texture, slightly thickened rim with no horn. Gill filaments four; lower two arising direct from basal trunk, upper two with a common stalk and a sharp kink a short distance from the common stalk, particularly on the upper filament; all filaments held close together (Fig. 43). These two species can be distinguished as adults but not in the pupal stage.

Ecology. *S. angustipes* and *S. velutinum* have been recorded in small seasonally-flowing streams, small permanent streams and, occasionally, larger streams and rivers; they are generally, but not exclusively, distributed in lowland areas of Britain and Ireland. Earlier distribution records referring to *angustipes* (e.g. Davies 1968; Post 1981; Crosskey 1982, 1985b) may have included *velutinum* (see Taxonomic Notes on page 9). Both *angustipes* and *velutinum* last-instar larvae have been recorded from early spring until autumn; a maximum of three generations annually is probable; the spring generation of both species is most strongly represented in small streams (Bass, unpublished).

Simulium aureum Fries

Larva. Body length 6–8 mm, brown or grey, with inconspicuous banding. Head capsule ground-colour pale brown, eyebrow mark consists of a single small pigment spot; apotome standard marks dark brown with fuzzy outlines (Fig. 10A). Postgenal cleft square with sharp corners, its shape easy to see on the pale brown head capsule (Fig. 10B); ratio of cleft width to head width *ca.* 1:5. Conspicuous ventral papillae. Rectal organ with three simple lobes. Pupal gill histoblast with four coiled filaments; histoblast outline with sharply angled corner antero-ventrally (Fig. 9A).

Pupa. Cocoon length *ca.* 4.0–4.5 mm; cocoon woven with coarse silk, its anterior rim slightly thickened; no horn present. Gill filaments four, lower two arising direct from the basal trunk, upper two with a common stalk; top filament deviates strongly upward in a gentle curve, returning to lie close to the tips of the other filaments (Fig. 42).

Ecology. *S. aureum* is confined to very small streams; with few exceptions these are of low pH (range 3.5–6.0), typically draining open or slightly wooded heaths and moors. Consequently, *aureum* is largely restricted to the north and west in Britain and Ireland (Davies 1968; Crosskey 1991), with isolated records from the New Forest, Dorset heaths (Bass & Brockhouse 1990) and Stanmore Common, near London (Leonhardt 1985).

Simulium lineatum (Meigen), *Simulium pseudequinum* Séguy and *Simulium equinum* (Linnaeus)

Larvae. In all three species, body length 6–8 mm, colour greenish brown; when preserved, body consistently adopts a strongly curved shape with posterior hook circlet held in a ventral position. Head capsule pale yellow to amber; apotome marks well defined, uniform dark brown, with antero-lateral marks clear-edged and paired on each side (Fig. 13C). Postgenal cleft broad with ill-defined margin (Fig. 13D); ratio of maximum cleft width to head width *ca.* 1:3. Ventral papillae small and inconspicuous. Rectal organ with three simple lobes. Pupal gill histoblast containing convoluted, not coiled, finger-like processes (Fig. 13E).

These three species are only separable as pupae and adults, although last-instar larvae of *equinum* may be distinguished from *pseudequinum* and *lineatum* by dissection of the pupal histoblast (see page 19; also see the key to pupae, page 64). Jensen (1984; in press) describes characters for separating *equinum* from *lineatum*, but they have not proved reliable with larvae from England (M. Ladle, pers. comm.).

Pupae. In all three species, cocoon length ranges from *ca.* 3 mm in summer to *ca.* 5 mm in spring; cocoon evenly woven with fine silk, lifting away from the substratum anteriorly to form a tubular opening; anterior rim unthickened; no horn present (Fig. 34). Gill consists of six short finger-like processes arising from an inflated basal trunk; the two basal trunks form a collar at the cocoon entrance.

S. lineatum – gill lobes narrower than the inflated basal trunk and of equal diameter for their full length (Fig. 34B).

S. pseudequinum – gill lobes narrower than inflated basal trunk but constricted in diameter at their attachment point to the basal trunk (Fig. 34C).

S. equinum – gill lobes similar in diameter to the inflated basal trunk (Fig. 34A).

Ecology. *S. lineatum* is a species of rivers and largely absent from streams, although its distribution extends upstream in summer (Ladle et al. 1977); generally associated with weedy lowland rivers in Britain and Ireland, though surprisingly absent from southeast England (Crosskey 1985b). Pupae occur in all months from March to November, with up to four generations per year recorded (Ladle et al. 1977); more generations per year have been achieved in laboratory culture (Ham & Bianco 1984).

S. pseudequinum is currently reported from only a few lowland rivers in central England (Dorset Stour and Rivers Kennet, Thame, Cherwell and Great Ouse) (Crosskey 1981); also some additional Thames tributaries near Oxford (Williams 1991). This species probably remains overlooked because of its close similarity to *lineatum*.

S. equinum is another species characteristic of weedy rivers in lowland areas but also present in small streams; the most widely occurring of the *Wilhelmia*-group species, with records from Scotland (Zahar 1951; Maitland & Penney 1967; Davies 1968), Wales (Davies 1968) and Ireland (Frost 1942; Fahy 1972; Dowling et al. 1981; Bass 1990). In the south, up to four generations per year occur in large rivers; the overwintering generation is frequently absent from smaller streams (Ladle et al. 1972, 1977; Hansford 1978).

Simulium erythrocephalum (De Geer)

Larva. Body length ranging from *ca.* 5 mm in summer to *ca.* 7 mm in early spring; body shape on preservation may be curved with posterior hook circlet held ventrally, resembling larvae of the subgenus *Wilhelmia (lineatum, pseudequinum* and *equinum*); colour pale grey, red-brown banding conspicuous on spring larvae. A pair of small inconspicuous papillae dorso-laterally on the side of each body segment (Fig. 14E). Head capsule ground-colour uniform pale yellow (slightly darker in spring); standard apotome marks compressed laterally, antero-lateral marks close to narrow antero- and postero-median marks (Fig. 14A); apotome marks of variable density, dark brown in spring and pale brown in summer. Postgenal cleft poorly defined against head capsule ground-colour (Fig. 14B); ratios of maximum cleft width to head width range from 1:2.75 to 1:3.0. Ventral papillae occasionally conspicuous, intermediate in size between those of the *Nevermannia/ Eusimulium* subgenera and the *Simulium/Wilhelmia* subgenera. Rectal organ with three simple lobes, though subsidiary lobules have been recorded (Puri 1925). Pupal gill histoblast with six coiled filaments.

Pupa. Cocoon length 3.0–4.5 mm; cocoon evenly woven of fine silk, slightly thickened anterior rim; no horn present. Gill filaments six; paired filaments arising on very short common stalks from basal trunk; arranged in two planes, the central pair deviating horizontally, the top and bottom pair deviating vertically (Fig. 29).

Ecology. A species of lowland streams and rivers, *S. erythrocephalum* has been widely recorded in England (Doby & Saguez 1963; Davies 1968; Hansford 1978; Post 1982, 1983; Crosskey 1985b), with isolated records from Scotland (Davies 1968) and Ireland (Fahy 1972; Bass 1990); strongly associated with water of high conductivity and pH values above 7 (Grunewald 1972). A clear morphological variability between spring and summer larvae has been long recognised (Edwards 1920 – as "*argyreatum*" var. *sericatum* (by mis-identification)); recent work has shown seven instars in spring and six instars in summer generations, with up to five generations each year (Post 1982, 1983). Pupae have been recorded from March to November.

Simulium ornatum Meigen, *Simulium intermedium* Roubaud and *Simulium trifasciatum* Curtis

Larvae. In all three species, body length 6–8 mm, colour pale grey. Head capsule ground-colour uniform pale yellow or light brown, darker specimens with more pigmentation to rear of head capsule; standard apotome marks present but of variable intensity; antero-lateral marks a single smudge on each side (Fig. 14C). Postgenal cleft square with rounded corners and anterior margin; cleft poorly defined on pale head capsules (Fig. 14D); ratio of cleft width to head width *ca.* 1:3.5. Ventral papillae small and inconspicuous. Rectal organ with three simple lobes. Pupal gill histoblast with eight coiled filaments.

These three species cannot be separated in the larval stages owing to a high degree of variability, particularly in *ornatum*. This variation is probably exacerbated by the presence of a species-complex in both *ornatum* and *intermedium*.

Pupae. Characters common to all three species:- cocoon length 3.5–5.0 mm, smallest specimens found at highest water temperatures; cocoon rim thickened; no horn present (Fig. 26). Gill filaments eight; filaments branching in pairs from common stalks of variable lengths, generally the common stalks deviate in vertical plane (Fig. 26), though lower common stalk may deviate horizontally.

Character shared by *ornatum* and *intermedium*:- sparse microtubercles on thoracic cuticle, microtubercles rounded in profile (Fig. 25B).

Character specific to *trifasciatum*:- dense microtubercles on thoracic cuticle, many with pointed apex in profile (Fig. 25A).

S. intermedium – cocoon loosely woven with coarse silk (Figs 22, 26B).

S. ornatum and *trifasciatum* – cocoon generally evenly woven with fine silk (Figs 22, 26A), but a few examples of *ornatum* cocoons with coarse silk have been found.

Ecology. Within the *S. ornatum* group, *trifasciatum* appears to have the most clear-cut ecological requirements, occurring in spring and autumn in very small streams; however, in southern England it is strongly associated with calcareous spring sources (Crosskey 1982, 1985b). By contrast, in small upland streams, base-poor conditions are preferred (Davies 1968). Crosskey (1991) records its presence in Ireland (see following note on species complexes).

S. intermedium has been recorded from streams and occasionally rivers (Maitland & Penney 1967; Davies 1968; Wotton 1976 – as *nitidifrons*). In southern England it is confined to streams draining heathland such as Dartmoor, the New Forest and East Sussex Weald (Davies 1968; Crosskey 1985b; Bass & Brockhouse 1990). In Ireland it is widely distributed (Fahy 1972 – as *nitidifrons*). Pupae may be found from spring to autumn.

S. ornatum has been recorded from most types of flowing water throughout much of Britain and Ireland from early spring until autumn. Its apparent plasticity may be explained by the discovery of four sibling species, currently only distinguished by cytological techniques (Post 1980); their specific habitat requirements and voltinism is unknown.

[NOTE – the presence of sibling species within *ornatum* (4) and *intermedium* (2) (Post 1980) exacerbates the problems associated with the recognition of "species" and defining their ecological characteristics; the contradictions mentioned in habitat requirements for *trifasciatum* may indicate that this is also a species-complex (an appropriate term!)].

Simulium argyreatum Meigen and *Simulium variegatum* Meigen

Larvae. In both species, body length 7–8 mm, colour slate grey with a contrasting off-white area ventrally towards hind end. Head capsule pale brown ventrally, shading to cream or pale yellow on the apotome (note: winter larvae have darker head capsules); standard apotome marks sometimes present, often poorly defined or restricted to a postero-median mark of variable density (Fig. 15A). Postgenal cleft well defined, tapering to a point anteriorly (Fig. 15B); cleft width generally less than one-third of maximum head width. Ventral papillae small and inconspicuous. Rectal organ with secondary branches (lobules). Pupal gill histoblast with six coiled filaments.

These two species cannot be separated in the larval stages. The comparative size of the postgenal cleft and density of head capsule ground-colour, previously used to distinguish these species, has proved unreliable when tested between single-species populations.

Pupae. In both species, cocoon length 3.5–4.5 mm; cocoon with no horn present. Gill filaments six, filaments rather short, three pairs on very short common stalks arising in sequence from a thicker basal trunk; filaments held close together, upper filaments thicker than lower filaments (Figs 28, 32).

S. variegatum – a pair of small oval lumps (patagia) at the front of thorax dorsally. Cocoon of coarse silk with an open weave anteriorly; anterior rim unthickened (Fig. 28).

S. argyreatum – patagia absent. Cocoon silk fine and evenly woven; anterior rim thickened (Fig. 32).

Ecology. *S. argyreatum* and *S. variegatum* are both common but restricted to the north and west of Britain (Edwards 1920; Zahar 1951; Maitland & Penney 1967; Davies 1968; Wotton 1976; Hywel-Jones & Ladle 1986) and are widely distributed in Ireland (Fahy 1972). They occur together in small to large turbulent stony streams, with *variegatum* extending downstream into large rivers with boulders. Life cycles are similar, with two or three annual generations in the north (Davies & Williams 1962) and three in the southwest (Hywel-Jones & Ladle 1986).

Simulium tuberosum (Lundström) species-complex

Larva. Body length 6–7 mm, colour grey to light brown. Head capsule ground-colour light yellow; pigmentation restricted to pale eyebrow marks and some lightly pigmented areas in antero-dorsal regions of the apotome and head capsule sides; paired eyespots equal in size (Fig. 15C). Postgenal cleft poorly defined with wide rounded anterior margin (Fig. 15D); maximum ratios of cleft width to head width range from *ca.* 1:2.2 to 1:2.5. Ventral papillae small and inconspicuous. Rectal organ with secondary branches (lobules). Pupal gill histoblast with six coiled filaments.

Pupae. Cocoon length *ca.* 3.5 mm; cocoon evenly woven from fine silk; anterior rim slightly thickened; no horn present. Pupa not completely covered by cocoon. Gill filaments six; filaments short and thin, held close together, paired, with short common stalks arising in sequence from a basal trunk of similar diameter (Fig. 33).

Ecology. *S. tuberosum* has been recorded from the north and west of Britain in stony rivers and large streams (Edwards 1920; Davies 1968; Wotton 1976). Its presence in Ireland is apparently uncertain (Crosskey 1991). One or two annual generations in early and mid summer, overwintering in the egg stage (Davies 1968).

Simulium rostratum (Lundström)

Larva. Body length 5–7 mm, colour grey brown. Head capsule ground-colour a uniform light brown, standard apotome marks expressed as small pale spots; eye spots comparatively small (Fig. 16C). Postgenal cleft rounded anteriorly, clearly visible against head capsule ground-colour (Fig. 16D); ratios of cleft width to head width range from 1:3.5 to 1:4.0. Ventral papillae small and inconspicuous. Rectal organ with secondary branches (lobules). Pupal gill histoblast with six coiled filaments.

Pupa. Cocoon length about 4 mm; cocoon thin, woven of fine silk, anterior rim unthickened; no horn present. Gill filaments six; thin filaments arise in pairs on short common stalks; filaments deviate in the vertical plane and remain apart at their tips (Fig. 30).

Ecology. *S. rostratum* is characteristic of natural pond and lake outlets, occurring in decreasing numbers for a few kilometres downstream of outlets; apparently confined to localised populations in northwest Britain (Davies 1968) and Ireland (Schröder 1988). Overwintering as eggs with one or two generations in mid and late summer (Davies 1968).

Simulium morsitans Edwards

Larva. Body length 5–7 mm, colour brown. Head capsule ground-colour pale yellow; apotome with a narrow pale brown "H"-shaped central mark; paired eyespots equal in size (Fig. 17C). Postgenal cleft of rounded outline, margin poorly defined by the pale head capsule (Fig. 17D); ratios of maximum cleft width to head width generally less than about 1:3. Ventral papillae small and inconspicuous. Rectal organ with secondary branches (lobules). Pupal gill histoblast with eight coiled filaments.

Pupa. Cocoon length *ca.* 3.5 mm; robust and closely woven (Fig. 22), anterior rim slightly thickened, no horn present. Gill filaments eight; paired filaments arise from four common stalks, the two lower common stalks deviating away from the side of pupa, whilst the two upper common stalks are directed anteriorly (Fig. 24).

Ecology. *S. morsitans* is a rare species of large weedy streams and rivers (Davies 1968); early records indicated two generations: one in early summer and a second in mid summer (Edwards 1920). Recently found at four sites on the River Forth (Scotland) and single sites on the rivers Teifi (Tyfi) (Wales), Yorkshire Derwent and By Brook (data from the Institute of Freshwater Ecology – River Communities Project). Currently unrecorded from Ireland (Crosskey 1991).

Simulium posticatum Meigen

Larva. Body length 5–7 mm, colour becoming yellow soon after preservation. Head capsule ground-colour pale yellow, eyebrow mark absent; single apotome mark a central inverted "U" (Fig. 17A). Postgenal cleft rounded, poorly defined against pale head capsule (Fig. 17B); ratios of maximum cleft width to head width range from 1:2.5 to 1:3.0. Ventral papillae small and inconspicuous. Rectal organ with secondary branches (lobules). Pupal gill histoblast with six coiled filaments.

Pupa. Cocoon length 3.5–4.0 mm; cocoon robust and thickly woven (Fig. 23), anterior rim slightly thickened; no horn present. Gill filaments six; paired filaments diverge in the vertical plane then return to lie close together at their tips (Fig. 31).

Ecology. *S. posticatum* occurs in central and southern England, generally in slow-flowing weedy rivers; occasionally abundant, with adults providing a seasonal human-biting problem. Colloquially known as the "Blandford Fly". Though not restricted to the River Stour in Dorset, a particularly large population is present downstream from the town of Blandford in April and May (Hansford 1978). An overspill population appears spasmodically in the Moors River and certain New Forest streams on the Hampshire/Dorset border. It has been recorded from the River Thames and some tributaries near Oxford (Williams 1991), a tributary of the River Soar, Leicestershire (Cassella & Hay 1991), the River Arrow in the Midlands, and the River Axe on the Devon/Dorset border. Older records (as *austeni*) indicate a similarly restricted range (Davies 1968). In Europe it is widespread (Zwick 1978) and has been found recently in northern Germany after being unrecorded for fifty years (Timm & Piper 1985). *S. posticatum* has a single spring generation, larvae hatching in February to April. The oviposition site, within bankside soil above normal water level, is particularly unusual in the Simuliidae (Ladle et al. 1985). Currently unrecorded from Ireland (Crosskey 1991).

Simulium reptans (Linnaeus)

Larva. Body length 5–7 mm, colour pale with light brown banding. Head capsule ground-colour pale yellow, eyebrow mark absent; paired eyespots of unequal size: posterior spots about ×2 anterior spots. Apotome* pale yellow, variably expressed mark on apotome posterior boundary and pale anterolateral marks on some individuals (Fig. 15E). Postgenal cleft rounded, expanding anteriorly with boundary poorly defined against pale head capsule (Fig. 15F); ratios of cleft maximum width to head width range from 1:2.5 to 1:3.0. Ventral papillae small and inconspicuous. Rectal organ with short secondary branches (lobules). Pupal gill histoblast with eight coiled filaments.

*A form with a dark central apotome mark (Fig. 17E) occurs in varying proportions in some populations in England (var. *galeratum* of Davies 1968).

Pupa. Cocoon length 3.5–4.0 mm; cocoon densely woven (Fig. 22) with anterior rim thickened, gaps in cocoon sides near the anterior rim; no horn present. Gill filaments eight; paired filaments arise from short common stalks (Fig. 21).

Ecology. *S. reptans* has been reported from northwest Britain and Ireland in stony rivers and large streams (Pentelow 1935; Maitland & Penney 1967; Davies 1968; Fahy 1972; Dowling et al. 1981), also from Dartmoor (Davies 1968). Restricted populations occur in the Rivers Wey (Surrey) and Rother (West Sussex) (Crosskey, pers. comm.). Two or three annual generations have been recorded. Where two generations occur, they are in mid and late summer (Zahar 1951; Davies 1968; Zwick 1974); a third spring generation has been reported (Smart 1944; Maitland & Penney 1967). Recent studies on adult biting activity in Wales indicated that two or three generations were present (P. J. McCall, pers. comm.).

Simulium noelleri Friederichs

Larva. Body length 6–8 mm, colour pale grey-brown. Head capsule ground-colour pale brown, conspicuous eyebrow mark; apotome heavily shaded incorporating an ill-defined "H"-shaped mark (Fig. 16A). Postgenal cleft deep, clear edged, tapering to a point (Fig. 16B); ratios of maximum cleft width to head width range from 1:3.25 to 1:3.5. Ventral papillae small and inconspicuous. Rectal organ with secondary branches (lobules). Pupal gill histoblast with eight coiled filaments, branching 3:3:2 or 3:2:1:2.

Pupa. Cocoon length 4–5 mm; most of cocoon loosely constructed with coarse silk (Fig. 22); no horn present. Gill filaments eight; with two groups of three and one pair of filaments (Fig. 20) or a group of three, two pairs and a single filament.

Ecology. *S. noelleri* occurs at outlets from ponds, lakes and reservoirs; frequently found at very high population densities for a short distance downstream. Extensive studies on larval ecology (e.g. Wotton 1982, 1984, 1985) have established that *noelleri* is capable of recovering suspended particles down to colloidal size and lives in much closer proximity to its nearest neighbours than is tolerated by other simuliids. The readiness with which *noelleri* colonises artificial outlets is well illustrated by its transient appearance in small experimental channels receiving pumped lakewater from Windermere, the channels being situated several miles from the nearest known population of *noelleri* (Bass, unpublished). A study on the frequency of occurrence of simuliids below fifty British reservoirs showed that *noelleri* was the only species displaying an affinity for such highly modified streams (Bass & Armitage 1987). *S. noelleri* is widely distributed throughout Britain and Ireland (as *argyreatum* in Davies 1968; Fahy 1972). After the comparatively synchronised emergence of an overwintering generation, a continuous presence of larvae and pupae can obscure the number of annual generations produced, though *noelleri* is clearly multivoltine in the British Isles.

ACKNOWLEDGEMENTS

I would like to thank Dr Mike Ladle for introducing me to the Simuliidae and also colleagues who provided preserved specimens, notably Dr James Copeland (*tuberosum* and *reptans* from Scotland), Dr Frank Jensen and Mr Rick Gunn (*morsitans* from Denmark and England, respectively). Dr Charles Brockhouse and the late Professor Klaus Rothfels inspired some comprehension of chromosome banding patterns. Advice and encouragement on the structure and contents of the keys from Dr Roger Crosskey and Carolyn Lowry (Natural History Museum) is gratefully acknowledged. The long-term support received from the Freshwater Biological Association, the Institute of Freshwater Ecology, and my wife, Lin Baldock, ensured that the task was finally completed.

REFERENCES

Bass, J. A. B. (1985). New keys to the immatures of British simuliids: plans, pitfalls and progress so far with the Eusimuliums. *Newsl. Brit. Simulium Group* **11**, 4-6.

Bass, J. (1990). Some records of black flies (Diptera: Simuliidae) from Co. Kerry. *Ir. Nat. J.* **23**, 305-309.

Bass, J. A. B. & Armitage, P. D. (1987). Observed and predicted occurrence of blackflies (Diptera: Simuliidae) at fifty reservoir outlets in Britain. *Reg. Rivers: Res. & Man.* **1**, 247-255.

Bass, J. A. B. & Brockhouse, C. (1990). A new British species of the *Simulium vernum* group, with comments on its ecology and life history (Diptera: Simuliidae). *Aquatic Insects* **12**, 65-84.

Bass, J. A. B., Crosskey, R. W. & Werner, D. (1995). On the European blackfly *Simulium lundstromi* and inclusion within this species of *S. latigonium* as a new synonym. *Bull. Brit. Simulium Group* **5**, 7-19.

Brockhouse, C. (1985). Sibling species and sex chromosomes in *Eusimulium vernum* (Diptera: Simuliidae). *Can. J. Zool.* **63**, 2145-2161.

Carlsson, M., Nilsson, L. M., Svensson, Bj., Ulfstrand, S. & Wotton, R. S. (1977). Lacustrine seston and other factors influencing the blackflies (Diptera: Simuliidae) inhabiting lake outlets in Swedish Lapland. *Oikos* **29**, 229-238.

Cassella, J. P. & Hay, J. (1991). Dermal lesions and *Simulium posticatum:* a report from central England. *The Entomologist* **110**, 29-32.

Crosskey, R. W. (1981). The identity and synonymy of *Simulium (Wilhelmia) pseudequinum* Séguy and the occurrence of this species in England (Diptera: Simuliidae). *Entomologist's Gaz.* **32**, 137-148.

Crosskey, R. W. (1982). The blackfly fauna of the Isle of Wight (Diptera: Simuliidae). *Entomologist's Gaz.* **33**, 199-212.

Crosskey, R. W. (1985a). The rediscovery of *Simulium yerberyi* Edwards in Britain and its part in a reassessment of variability in *Simulium (Hellichiella) latipes* (Meigen) (Diptera: Simuliidae). *Entomologist's Gaz.* **36**, 209-226.

Crosskey, R. W. (1985b). The Blackfly fauna of the London area (Diptera: Simuliidae). *Entomologist's Gaz.* **36**, 55-75.

Crosskey, R. W. (1987a). An annotated checklist of the world black flies (Diptera: Simuliidae). In *Black Flies: Ecology, Population Management and Annotated World List* (eds K. C. Kim & R. W. Merritt), pp. 425-520. Pennsylvania State University, Pennsylvania.

Crosskey, R. W. (1987b). The blackfly fauna of Madeira (Diptera: Simuliidae). *Entomologist's Gaz.* **38**, 143-157.

Crosskey, R. W. (1990). *The Natural History of Blackflies.* John Wiley & Sons, Chichester. 711 pp.

Crosskey, R. W. (1991). A new checklist of the blackflies of Britain and Ireland, with geographical and type information (Diptera: Simuliidae). *Entomologist's Gaz.* **42**, 206-217.

Crosskey, R. W. & Davies, L. (1972). The identity of *Simulium lineatum* (Meigen), *S. latipes* (Meigen) and *S. vernum* (Macquart) (Diptera: Simuliidae). *Entomologist's Gaz.* **23**, 249-258.

Davies, L. (1957). A new *Prosimulium* species from Britain and a re-examination of *P. hirtipes* Fries from the Holarctic region (Diptera, Simuliidae). *Proc. R. ent. Soc. London (B)* **26**, 1-10.

Davies, L. (1966). The taxonomy of British black-flies (Diptera: Simuliidae). *Trans. R. ent. Soc. Lond.* **118**, 413-511.

Davies, L. (1968). A key to the British species of Simuliidae (Diptera) in the larval, pupal and adult stages. *Scient. Publs Freshwat. Biol. Ass.* **24**, 1-126.

Davies, L. & Smith, C. D. (1958). The distribution and growth of *Prosimulium* larvae (Diptera: Simuliidae) in hill streams in northern England. *J. Anim. Ecol.* **27**, 335-348.

Davies, L. & Williams, C. B. (1962). Studies on black flies (Diptera: Simuliidae) taken in a light trap in Scotland. 1. Seasonal distribution, sex ratio and internal condition of catches. *Trans. R. ent. Soc. Lond.* **114**, 1-20.

Doby, J.-M. & Saguez, F. (1963). Répartition géographique comparée des espèces des Simulies des Groupes *monticola* et *latipes* dans le Sud de la Grande-Bretagne et en Bretagne française. *Bull. Soc. Pharmacie de l'Quest* **3**, 69-78.

Dowling, C., O'Connor, J. P. & O'Grady, M. F. (1981). A baseline survey of the Caragh, an unpolluted river in southwest Ireland: observations on the macroinvertebrates. *J. Life Sci. R. Dubl. Soc.* **2**, 147-159.

Dunbar, R. W. (1959). The salivary gland chromosomes of seven forms of black flies included in *Eusimulium aureum* Fries. *Can. J. Zool.* **37**, 495-525.

Edwards, F. W. (1920). On the British species of *Simulium*. II. The early stages; with corrections and additions to Part I. *Bulletin Ent. Res.* **11**, 211-246.

Edwards, F. W. (1927). Notes on British *Simulium* (Diptera). *Ent. mon. Mag., London* **63**, 255-257.

Fahy, E. (1972). A preliminary account of the Simuliidae (Diptera) in Ireland, with observations on the growth of three species. *Proc. R. Ir. Acad.* **72B**, 75-81.

Frost, W. E. (1942). R. Liffey Survey IV. The fauna of the submerged "mosses" in an acid and an alkaline water. *Proc. R. Ir. Acad.* **47B**, 293-369.

Grenier, P. (1947). Notes morphologiques et biologiques sur quelque Simulies nouvelles pour la faune française. *Bull. Soc. Ent. France, Paris* **52**, 66-69.

Grunewald, J. (1965). Zür Kenntnis der simuliidenfauna (Diptera) des Sud-Schwarzwaldes und seiner Randgebeite. *Beitr. Naturk. Forsüch. SW-Deutchl.* **24**, 143-152.

Grunewald, J. (1972). Die Hydrochemischen Leibensbedingungen der priamaginalen Stadien von *Boophthora erythrocephala* De Geer (Diptera: Simuliidae). 1. Freilanduntersuchungen. *Zeitschrift für Tropenmedizin und Parasitologie* **24**, 432-445.

Ham, P. J. & Bianco, A. E. (1984). Maintenance of *Simulium (Wilhelmia) lineatum* Meigen and *Simulium erythrocephalum* De Geer through successive generations in the laboratory. *Can. J. Zool.* **62**, 870-877.

Hansford, R. G. (1978). Life history and distribution of *Simulium austeni* (Diptera: Simuliidae) in relation to phytoplankton in some English rivers. *Freshwat. Biol.* **8**, 521-531.

Hywel-Jones, N. L. & Ladle, M. (1986). Ovipositional behaviour of *Simulium argyreatum* and *S. variegatum* and its relationship to infection by the fungus *Erynia conica* (Entomophthoraceae). *Freshwat. Biol.* **16**, 397-403.

Jensen, F. (1984). A revision of the taxonomy and distribution of the Danish blackflies (Diptera: Simuliidae), with keys to the larval and pupal stages. *Natura Jutlandica* **21**, 69-116.

Knoz, J. (1965). To identification of Czechoslovakian Black-flies (Diptera, Simuliidae). *Folia Prirod. Kak. Univ. I.E. Purkyne (Biol. 2), Brno* **6**, 1-54 + 425 Abb.

Ladle, M. & Bass, J. A. B. (1975). A new species of *Metacnephia* Crosskey (Diptera: Simuliidae) from the south of England, with notes on its habitat and biology. *Hydrobiologia* **47**, 193-207.

Ladle, M. & Esmat, A. (1972). Records of Simuliidae (Diptera) from the Bere Stream, Dorset, with details of the life history and larval growth of *Simulium (Eusimulium) latipes* Meigen. *Entomologist's Monthly Mag.* **108**, 167-172.

Ladle, M., Bass, J. A. B. & Cannicott, L. J. (1985). A unique strategy of blackfly oviposition (Diptera: Simuliidae). *Entomologist's Gaz.* **36**, 147-149.

Ladle, M., Bass, J. A. B. & Jenkins, W. R. (1972). Studies on production and food consumption by the larval Simuliidae of a chalk stream. *Hydrobiologia* **39**, 429-448.

Ladle, M., Bass, J. A. B., Philpott, F. & Jeffery, A. (1977). Observations on the ecology of Simuliidae from the R. Frome, Dorset. *Ecological Entomol.* **2**, 197-204.

Leonhardt, K. G. (1985). A cytological study of species in the *Eusimulium aureum* group (Diptera: Simuliidae). *Can. J. Zool.* **63**, 2043-2061.

Maitland, P. S. & Penney, M. M. (1967). The ecology of the Simuliidae in a Scottish river. *J. Anim. Ecol.* **36**, 178-206.

Niesiolowski, S. (1971). Biologia meszki *Eusimulium costatum* (Friederichs) (Simuliidae: Diptera). *Polskie Pismo Entomologiczne Bulletin Entomologique de Pologne* **41**, 161-168.

Pentelow, F. T. K. (1935). Notes on the distribution of the larvae and pupae of Simuliidae spp. in the R. Tees and its tributaries. *Parasitology* **27**, 543-546.

Post, R. J. (1980). Cytotaxonomy of the *Simulium ornatum* species-group in Britain. *Newsl. Brit. Simulium Group* **3**, 3-5.

Post, R. J. (1981). The distribution of blackflies (Diptera: Simuliidae) in Norfolk. *Trans. Norfolk Nat. Soc.* **25**, 153-163.

Post, R. J. (1982). Notes on the natural history of *Simulium (Boophthora) erythrocephalum* de Geer (Diptera: Simuliidae). *Entomologist's Monthly Mag.* **113**, 31-35.

Post, R. J. (1983). The annual cycle of *Simulium erythrocephalum* (Diptera: Simuliidae) at a site in Norfolk. *Freshwat. Biol.* **13**, 379-388.

Procunier, W. S. (1982). A cytological description of 10 taxa in *Metacnephia* (Diptera: Simuliidae). *Can. J. Zool.* **60**, 2852-2865.

Puri, I. M. (1925). On the life-history and structure of the early stages of Simuliidae (Diptera: Nematocera). Part II. *Parasitology* **17**, 335-369.

Raastad, J. E. (1975). Fordeling av knott (Diptera: Simuliidae) i Berbyvassdraget, Iddi Østfold. Nytt Fra Universitets Zoologiske Museum Oslo NR50. *Fauna* **28**, 92-96.

Schröder, P. (1988). Distribution patterns of blackfly (Diptera: Simuliidae) associations in two Irish river systems. *Hydrobiologia* **164**, 149-160.

Smart, J. (1944). The British Simuliidae with keys to the species in the adult, pupal and larval stages. *Scient. Publ. Freshwat. Biol. Ass.* **9**. 57 pp.

Timm, T. & Piper, W. (1985). *Simulium posticatum* Meigen, 1838, die "Blandford-Mücke" in Norddeutschland (Diptera: Simuliidae). *Entomologische Mitteilungen aus dem Zoologischen Museum, Hamburg* **125**, 109-117.

Williams, T. (1991). The ubiquitous occurrence of *Simulium posticatum* (Diptera) in rivers around Oxford. *Entomologist* **110**, 33-36.

Williams, T. R. (1980). A British simuliid with microsculptured larval cuticle. *Newsl. Brit. Simulium Group* **4**, 3-4.

Wotton, R. S. (1976). The distribution of blackfly larvae (Diptera: Simuliidae) in Upper Teesdale streams, northern England. *Hydrobiologia* **51**, 259-263.

Wotton, R. S. (1982). Does the surface film of lakes provide a source of food for animals living in the lake outlets? *Limnol. Oceanogr.* **27**, 959-960.

Wotton, R. S. (1984). The relationship between food particle size and larval size in *Simulium noelleri* Friederichs. *Freshwat. Biol.* **14**, 547-550.

Wotton, R. S. (1985). The reaction of larvae of *Simulium noelleri* (Diptera) to different current velocities. *Hydrobiologia* **123**, 215-218.

Zahar, A. R. (1951). The ecology and distribution of black-flies in S.E. Scotland. *J. Anim. Ecol.* **20**, 33-62.

Zwick, H. (1974). Faunistisch-ökologische und taxonomische Untersuchungen an Simuliidae (Diptera), unter besonderer Berücksichtigung der Arten des Fulda-Gebietes. *Abh. Senckenb. Naturforsch. Ges.* **533**, 1-116.

Zwick, H. (1978). Simuliidae. In *Limnofauna Europaea* (ed. J. Illies), pp. 396-403. Swets & Zeitlinger, Amsterdam.

Zwick, H. & Crosskey, R. W. (1980). The taxonomy and nomenclature of the blackflies (Diptera: Simuliidae) described by J. W. Meigen. *Aquatic Insects* **2**, 225-247.

Zwick, H. & Zwick, P. (1990). Terrestrial mass-oviposition of *Prosimulium*-species (Diptera: Simuliidae). *Aquatic Insects* **12**, 33-46.

INDEX TO SPECIES

Page numbers in **bold** type indicate illustrations

Boophthora 7

Cnephia tredecimatum 7

Eusimulium 6, 89

Hellichiella 6

Metacnephia 6-7
  amphora 5-7, 34, **35**, 45, 48, **49**, **52**, 78
  tredecimata 7

Nevermannia 6, 89

Prosimulium 6, **21**, 44, 76
  arvernense 6-7, 78
  hirtipes 6-7, 22, **23**, 45, 48, **49**, 76-7
  inflatum 6-7, 77
  latimucro 6-7, 22, **23**, 45, 48, **49**, 77
  tomosvaryi 6-7, 22, **23**, 45, **47**, 77-8

Simulium
  angustipes 6, 9, 28, **29**, 45, 72, 73, 87
  angustitarse 6, 8, 26, **27**, 45, **53**, 65, 74, **75**, 79-80
  argyreatum 7, 10, 38, **39**, 45, **62**, 90-2, 96
  armoricanum 6, 32, **33**, 45, **68-9**, 80, 83
  aureum 6, **28-9**, 45, 72, **73**, 87-8
  austeni 7, 10, 94
  brevicaule 6, 8, 84
  cambriense 6, 8, 80
  carthusiense 8
  costatum 6, 26, **27**, 45, **53**, **65**, 86
  crenobium 85
  cryophilum 6, 8, 30, **31**, 45, **69**, 70, **71**, 80, 83-5
  dunfellense 6, 32, **33**, 45, 70, **71**, 80, 83-6
  equinum 6, 10, 34, **35**, 45, **64**, 88-9

  erythrocephalum 7, 24, 34, 36, **37**, 45, **59**, 89-90
  intermedium 7, 10, 36, 45, **52**, **55-6**, 90-1
  juxtacrenobium 5-6, 9, 30, **31**, 45, **53**, 66, **67**, 84-5
  latigonium 6, 8
  latipes 6, 8, **21**, 24, **25**, 45, **57**, 79-81, 84
  lineatum 6, 9, 34, **35**, 45, **64**, 88-9
  lundstromi 6, 8, 26, **27**, 45, **53**, 74, **75**, 80
  monticola 7, 10
  morsitans 7, 42, **43**, 45, **52**, **54**, 93-4
  naturale 6, 8-9, 81-2
  nitidifrons 7, 10, 91
  noelleri 7, 10, 40, **41**, 45, **50**, **52**, 95, 96
  ornatum 7, 10, 36, **37**, 45, **52**, **55-6**, 90-1
  posticatum 7, 10, 42, **43**, 45, **53**, 60, **61**, 94
  pseudequinum 5-6, 9, 34, **35**, 45, **64**, 88-9
  reptans 7, 38, **39**, 42, **43**, 45, **51-2**, 95
  rostratum 7, 10, 40, **41**, 45, **60**, 93
  salopiense 6, 9
  spinosum 7
  subexcisum 6, 8, 79
  sublacustre 7, 10
  trifasciatum 7, 10, 36, 45, **52**, **55-6**, 90-1
  tuberosum 7, 10, 38, **39**, 45, **63**, 92-3
  urbanum 6, 32, **33**, 45, 70, **71**, 80, 83-6
  variegatum 7, 38, **39**, 45, **58**, 91-2
  velutinum 6, 9, 28, **29**, 45, 72, **73**, 87
  vernum 6, 8-9, 30, **31**, 45, **53**, 66, **67**, **69**, 80-2, 84-5
  yerberyi 57
  zetlandense 6, 10

Wilhelmia 6, 89

PUBLICATIONS OF THE FRESHWATER BIOLOGICAL ASSOCIATION

The publications listed on the following pages are some of those which are currently available for purchase. Order forms and further information may be obtained from:- **Dept. DWS, Freshwater Biological Association, The Ferry House, Far Sawrey, Ambleside, Cumbria LA22 0LP, UK.** The FBA regrets that it is unable to send books on approval.

SCIENTIFIC PUBLICATIONS (SPs)

SP 5. A KEY TO THE BRITISH SPECIES OF FRESHWATER CLADOCERA, WITH NOTES ON THEIR ECOLOGY, by D.J. Scourfield & J.P. Harding.
Third edition, 1966. (Reprinted 1994). Pp. 1-61. ISBN 0 900386 01 0.

SP 13. A KEY TO THE BRITISH FRESH- AND BRACKISH-WATER GASTROPODS, WITH NOTES ON THEIR ECOLOGY, by T.T. Macan.
Fourth edition, 1977. (Reprinted 1994). Pp. 1-46. ISBN 0 900386 30 4.

SP 17. A KEY TO THE ADULTS AND NYMPHS OF THE BRITISH STONEFLIES (PLECOPTERA), WITH NOTES ON THEIR ECOLOGY AND DISTRIBUTION,
by H.B.N. Hynes. Third edition, 1977. (Reprinted 1993). Pp. 1-92. ISBN 0 900386 28 2.

SP 18. A KEY TO THE BRITISH FRESHWATER CYCLOPID AND CALANOID COPEPODS, WITH ECOLOGICAL NOTES, by J.P. Harding & W.A. Smith.
Second edition, 1974. Pp. 1-56. ISBN 0 900386 20 7.

SP 23. A KEY TO THE BRITISH SPECIES OF FRESHWATER TRICLADS (TURBELLARIA, PALUDICOLA), by T.B. Reynoldson.
Second edition, 1978. Pp. 1-32 + 1 colour plate. ISBN 0 900386 34 7.

SP 25. SOME METHODS FOR THE STATISTICAL ANALYSIS OF SAMPLES OF BENTHIC INVERTEBRATES, by J.M. Elliott.
Second edition, 1977. (Reprinted 1993). Pp. 1-160. ISBN 0 900386 29 0.

SP 27. A KEY TO THE FRESHWATER FISHES OF THE BRITISH ISLES, WITH NOTES ON THEIR DISTRIBUTION AND ECOLOGY, by P.S. Maitland.
1972. Pp. 1-139. ISBN 0 900386 18 5.

SP 29. TURBULENCE IN LAKES AND RIVERS, by I.R. Smith. 1975. Pp. 1-79. ISBN 0 900386 21 5.

SP 30. AN ILLUSTRATED GUIDE TO AQUATIC AND WATER-BORNE HYPHOMYCETES (FUNGI IMPERFECTI), WITH NOTES ON THEIR BIOLOGY,
by C.T. Ingold. 1975. Pp. 1-96. ISBN 0 900386 22 3.

SP 31. A KEY TO THE LARVAE, PUPAE AND ADULTS OF THE BRITISH DIXIDAE (DIPTERA), THE MENISCUS MIDGES, by R.H.L. Disney.
1975. Pp. 1-78. ISBN 0 900386 23 1.

SP 33. DEPTH CHARTS OF THE CUMBRIAN LAKES by A.E. Ramsbottom. 1976. Pp. 1-39. ISBN 0 900386 25 8. [Data are given for 15 lakes.]

FBA PUBLICATIONS

SP 34. AN ILLUSTRATED KEY TO FRESHWATER AND SOIL AMOEBAE, WITH NOTES ON CULTIVATION AND ECOLOGY, by F.C. Page. 1976. Pp. 1-155. ISBN 0 900386 26 6.

SP 36. WATER ANALYSIS: SOME REVISED METHODS FOR LIMNOLOGISTS, by F.J.H. Mackereth, J. Heron & J.F. Talling. Second impression, 1989. Pp. 1-120. ISBN 0 90386 31 2.

SP 38. A KEY TO THE FRESHWATER PLANKTONIC AND SEMI-PLANKTONIC ROTIFERA OF THE BRITISH ISLES, by R.M. Pontin. 1978. Pp 1-178. ISBN 0 900386 33 9.

SP 39. A GUIDE TO METHODS FOR ESTIMATING MICROBIAL NUMBERS AND BIOMASS IN FRESH WATER, by J.G. Jones. 1979. Pp. 1-112. ISBN 0 900386 37 1.

SP 40. A KEY TO THE BRITISH FRESHWATER LEECHES, WITH NOTES ON THEIR LIFE CYCLES AND ECOLOGY, by J.M. Elliott & K.H. Mann. 1979. (Reprinted 1998). Pp. 1-72 + 1 colour plate. ISBN 0 900386 38 X.

SP 41. A KEY TO THE BRITISH AND EUROPEAN FRESHWATER BRYOZOANS, by S.P. Mundy. 1980. Pp. 1-31. ISBN 0 900386 39 8.

SP 42. A KEY TO THE COMMONER DESMIDS OF THE ENGLISH LAKE DISTRICT, by E.M. Lind & A.J. Brook. 1980. Pp. 1-123. ISBN 0 900386 40 1.

SP 44. A GUIDE TO THE MORPHOLOGY OF THE DIATOM FRUSTULE, WITH A KEY TO THE BRITISH FRESHWATER GENERA, by H.G. Barber & E.Y. Haworth. 1981. (Reprinted 1994). Pp. 1-112. ISBN 0 900386 42 8.

SP 45. A KEY TO THE LARVAE OF THE BRITISH ORTHOCLADIINAE (CHIRONOMIDAE), by P.S. Cranston. 1982. Pp. 1-152 + 1 plate. ISBN 0 900386 43 6.

SP 46. THE PARASITIC COPEPODA AND BRANCHIURA OF BRITISH FRESHWATER FISHES: A HANDBOOK AND KEY, by G. Fryer. 1982. Pp. 1-87. ISBN 0 900386 44 4.

SP 47. A KEY TO THE ADULTS OF THE BRITISH EPHEMEROPTERA, WITH NOTES ON THEIR ECOLOGY, by J.M. Elliott & U.H. Humpesch. 1983. Pp. 1-101 + 1 plate. ISBN 0 900386 45 2.

SP 48. KEYS TO THE ADULTS, MALE HYPOPYGIA, FOURTH-INSTAR LARVAE AND PUPAE OF THE BRITISH MOSQUITOES (CULICIDAE), WITH NOTES ON THEIR ECOLOGY AND MEDICAL IMPORTANCE, by P.S. Cranston, C.D. Ramsdale, K.R. Snow & G.B. White. 1987. Pp. 1-152. ISBN 0 900386 46 0.

SP 49. LARVAE OF THE BRITISH EPHEMEROPTERA: A KEY WITH ECOLOGICAL NOTES, by J.M. Elliott, U.H. Humpesch & T.T. Macan. 1988. Pp. 1-145. ISBN 0 900386 47 9.

SP 50. ADULTS OF THE BRITISH AQUATIC HEMIPTERA HETEROPTERA: A KEY WITH ECOLOGICAL NOTES, by A.A. Savage. 1989. Pp. 1-173. ISBN 0 900386 48 7.